Riverlands of the Anthropocene

This is an invitation to readers to ponder universal questions about human relations with rivers and water for the precarious times of the Anthropocene. The book asks how humans can learn through sensory embodied encounters with local waterways that shape the architecture of cities and make global connections with environments everywhere.

The book considers human becomings with urban waterways to address some of the major conceptual challenges of the Anthropocene, through stories of trauma and healing, environmental activism, and encounters with the living beings that inhabit waterways. Its unique contribution is to bring together Australian Aboriginal knowledges with contemporary western, new materialist, posthuman and Deleuzean philosophies, foregrounding how visual, creative and artistic forms can assist us in thinking beyond the constraints of western thought to enable other modes of being and knowing the world for an unpredictable future.

Riverlands of the Anthropocene will be of particular interest to those studying the Anthropocene through the lenses of environmental humanities, environmental education, philosophy, ecofeminism and cultural studies.

Margaret Somerville is Professor of Education and the Director of the Centre for Educational Research in the School of Education, University of Western Sydney, Australia.

Routledge Explorations in Environmental Studies

Romantic Anti-capitalism and Nature
The Enchanted Garden
Robert Sayre and Michael Löwy

Liberty and the Ecological Crisis
Freedom on a Finite Planet
Edited by Christopher J. Orr, Kaitlin Kish, and Bruce Jennings

Environmental Justice and Oil Pollution Laws
Comparing Enforcement in the United States and Nigeria
Eloamaka Carol Okonkwo

The Environmental Impact of Overpopulation
The Ethics of Procreation
Trevor Hedberg

Riverlands of the Anthropocene
Walking Our Waterways as Places of Becoming
Margaret Somerville

An Environmental History of Australian Rainforests until 1939
Fire, Rain, Settlers and Conservation
Warwick Frost

Daoism and Environmental Philosophy
Nourishing Life
Eric S. Nelson

https://www.routledge.com/Routledge-Explorations-in-Environmental-Studies/bookseries/REES

Riverlands of the Anthropocene

Walking Our Waterways as Places of Becoming

Margaret Somerville

Routledge
Taylor & Francis Group
LONDON AND NEW YORK

earthscan
from Routledge

First published 2020
by Routledge
2 Park Square, Milton Park, Abingdon, Oxon OX14 4RN

and by Routledge
52 Vanderbilt Avenue, New York, NY 10017

Routledge is an imprint of the Taylor & Francis Group, an informa business

Visit the eResources: www.routledge.com/9780815399964

British Library Cataloguing-in-Publication Data
A catalogue record for this book is available from the British Library

Library of Congress Cataloging-in-Publication Data
Names: Somerville, Margaret, author.
Title: Riverlands of the Anthropocene : walking our waterways as places of becoming / Margaret Somerville.
Description: Abingdon, Oxon ; New York : Routledge, 2020. |
Series: Routledge explorations in environmental studies |
Includes bibliographical references and index.
Identifiers: LCCN 2020002885 (print) | LCCN 2020002886 (ebook) |
ISBN 9780815399964 (hardback) | ISBN 9781351171120 (ebook)
Subjects: LCSH: Rivers--Environmental aspects. | Human ecology. |
Geology, Stratigraphic--Anthropocene.
Classification: LCC GF63 .S67 2020 (print) | LCC GF63 (ebook) |
DDC 304.20916/93--dc23
LC record available at https://lccn.loc.gov/2020002885
LC ebook record available at https://lccn.loc.gov/2020002886

ISBN: 978-0-8153-9996-4 (hbk)
ISBN: 978-1-351-17112-0 (ebk)

Typeset in Bembo
by Taylor & Francis Books

To our life-giving global waterways and all those who have looked after them since time immemorial, and the Gundungarra D'harawal, Darug, Guringai and Gadigal peoples of the Hawkesbury Nepean River system, who continue to care today.

Contents

	List of figures	viii
	Preface	ix
	Acknowledgements	xiv
1	A blue literature review	1
2	Rivers of the Anthropocene	25
3	Watery beginnings	46
4	Songlines: Walking the songlines of the singing painting river	65
5	Riverlands' watery ways	86
6	Bedrock's sacramental becomings	105
7	The river's crossings	124
8	Global materialities: The artful excess of river's litter	141
9	Regeneration: Of trees, weeds and tender intimacies	158
10	Life and death in the Anthropocene	176
Epilogue		194
	Index	200

Figures

1.1 Riverlands bower 2
1.2 The bower 3
3.1 Water and light 50
3.2 Wee Davey and son, Warragamba Dam construction 63
4.1 Pintupi Luritja women 66
4.2 The Humble Petitioner, ochre, collage, charcoal and acrylic, 2000 x1000 mm 71

Preface

The mattering of time and place in the riverlands

> ***Bedrock***
> At many times of day and season
> still surface
> of river is opaque
> at others it is lit
> by play of sky and light
> to reveal its bedrock.
> I know it by touch of hands
> as they feel their way
> along river's rocky bottom
> its ledges crevices
> loose stones sharp edges
> furry surfaces
> curved ridges when too weak
> to swim.

Riverlands is both a moment in time and all time. It is a sequel to *Water in a Dry Land,* continuing on from the end of this book and located in my move to the Nepean River in Sydney's water catchment. A place of belonging and not belonging, it is both new and gathers up all that has been before. It begins with a small extract from the last words of *Water in a Dry Land.* This writing, and the Preface itself, gesture towards many of the strands to be explored in *Riverlands.* It considers the time of a moment, the time of a human life, the deep time of Aboriginal onto-epistemologies, and the geological time of the Anthropocene in the place of the river's lands. It is the river's book.

The time and place of a (human) life

3 March 2012

As I come to the end of this writing I am in a new place, learning new country, country of the Darug peoples. I have left another homeplace, again a university refugee. I choose a house in the furthest corner of western Sydney at Emu Heights, said to be named by explorers for the emus that once grazed there. 'Welcome, Emu Girrl', the conveyancers say. The house sits between bush and river on the very edge of the fastest-growing region of Australia: 1.9 million people from more than 170 different countries, speaking over 100 languages, 35 per cent recently arrived from overseas. Feeling the global flows of people and place. I am also getting to know a new river.

Clambering through thick undergrowth just a few hundred metres from the new house, the Nepean River dances and bubbles as water meets rock and stone. Further upstream, out from Mulgoa, walking through sandstone country to The Rock Lookout, the river is silver serpent, winding through dark, tree-filled gorge thousands of metres below. I surf the river on the net to where it turns east and becomes the Hawkesbury River, following its wavery line of blue until it swells into the waters of Pittwater around Scotland Island where I had my first two babies. It enters the sea at Barrenjoey, Sydney's northernmost headland, just past where we camped with grandma and granddad in a big white canvas tent, small child-self sliding down huge white sandhills.

Last night, six hours ago, the waters from Warragamba Dam were released as the dam overflowed its capacity and the gates were opened. This morning instead of a quiet presence only audible from the sound of millions of frogs coming with the rains, the river roars. I listen in wonder; it is unmistakably the sound of the river, not the noise of traffic that sometimes blows on the wind, but a body sound of surging water, of movement and energy (Somerville, 2013, 190–191).

The time and place of the riverland's moments

> ***Riverlands, 10 May 2018, 7.20 a.m.***
> Shining white
> egret moment
> until pelican and white egret create
> another moment as
> pelican opens wide beak
> snap snap at egret
> egret moves quickly
> to another spot
> and guttural caw caws at pelican who
> splashes and floats off majestically
> down river to a roll of thunder
> as river darkens with cloud.
> It is only here I know who I am
> and river knows me.

This new book is written from a series of moments or encounters recorded in the riverlands over a period of six years from 2013–2019. It is the 'slow writing' that Deborah Bird Rose (2013, 2) proposes as 'one response to our impossible position as participants in, and witnesses to, catastrophes beyond our comprehension'. The moments are recorded in images, tiny videos and notes, in an echo of Deleuze's ruminations on time: 'A season, a winter, a summer, an hour, a date have a perfect individuality lacking nothing, even though this individuality is different from that of a thing or a subject' (Deleuze and Guattari, 1987, 261).

Each encounter is experienced as a process of becoming other to oneself: 'It is the wolf itself, and the horse, and the child, that cease to be subjects to become events, in assemblages that are inseparable from an hour, a season, an atmosphere, an air, a life' (ibid., 262). In the riverlands, 'Climate, wind, season, hour are not of another nature than the things, animals, or people that populate them, follow them, sleep and awaken within them' (ibid., 263). There are only 'affects, subjectless individuations that constitute collective assemblages. Nothing develops, but things arrive late or early, and form this or that assemblage depending on their compositions of speed' (ibid., 266). 'It is as though an immense plane of consistency of variable speed were forever sweeping up forms and functions, forms and subjects, extracting from them particles and affects. A clock keeping a whole assortment of times' (ibid., 271).

Deep time of Aboriginal onto-epistemologies of the riverlands

> ***Riverlands, 8 September 2018***
> Fold body into
> deep crevice
> bright blue of
> hovering dragonfly
> hum of cicada song
> soft calling of casuarinas
> in light breeze
> blowing rivers breath into mine
> and calling names of all who ever
> walked here shedding skin
> absorbed into substance of tree
> becoming tree.

The riverlands inhabit the (dream)time/space in the ever-present casuarinas or sheoaks that line its river's banks. I learn about the (dream)time of sheoaks from Noel Nannup, Nungar Elder who says:

> Every person is born in a catchment, so the rain falls, the water flows, so does spirit and they're all intertwined, and as we move across land, we shed from the largest organ of our body, which is our skin and it falls to the soil … the skin would be taken down into soil through mycorrhizals and the vascular bundles of the sheoak tree and our DNA would be in the tree. Those trees will only grow where there is close surface water and as it does that. it means the name we give those trees is *kwell*, the tree of our

> ancestors, the tree of names, the tree of the spirits, and on a windy day. you will hear the voices of every person that's ever lived and their skin has been shed and taken up by the tree and you will hear their names.
>
> (Noel Nannup, in Chinna, 2017)

I come to understand that '*if you are born here, the Country knows you. If you've lived here for more than six years, the Country knows you, and if you intend to stay, you have a responsibility and that is to care*' (ibid.). My search for belonging is over, I am here in this riverlands where I have lived now for six years in this new geological age of the Anthropocene.

The planet's time and place/s: the Anthropocene

> Always the blue of sky
> as body trawls through water immersed
> becoming water
> as hands feel their way along bedrock
> of river's intimacy
> curves crevices loose stones ridges
> always water rock stone and
> blue of sky entangled
> in sublime inhumanity.

The new geological epoch of the Anthropocene is described as the age of human entanglement in the fate of the planet, but what can this mean in relation to the time of the moment, the encounter, and in particular, what do the riverlands ask of the world in this time of the Anthropocene? For Barad,

> The encounter rests on an acknowledgement of an elemental otherness that is related to our own: We do not relate to the light, the earth, the air, and the warmth only with our individual sensibility and sensuality. We communicate to one another the light our eyes know, the ground that sustains our postures, and the air and the warmth with which we speak. We face one another as condensations of earth, light, air, and warmth, and orient one another in the elemental in a primary communication.
>
> (2012, 218)

She calls this 'a sublime inhumanity' (ibid., 219). Each moment in the riverlands reaches towards the entanglement of stone and rock, earth and water, and the sublime inhumanity of blue as the river flows within 'our blue planet', an interconnected matrix of rivers and oceans. My river is never just one river, but always already interconnected to the great rivers of the world.

References

Barad, K. (2012). On touching: the inhuman that therefore I am. *Differences: A Journal of Feminist Cultural Studies*, 23(5), 206–223.

Chinna, N. (2017). Life-giving water. *ABC Radio*. Available at: www.abc.net.au/radionational/programs/bigideas/life-giving-water/8356708

Deleuze, G. and Guattari, F. (1987). *A Thousand Plateaus: Capitalism and Schizophrenia*, trans. B. Massumi, Minneapolis: University of Minnesota Press.

Rose, D.B. (2013). Slowly ~ writing into the Anthropocene, ed. M. Harrison, D. Rose, L. Shannon and K. Satchell, Special Issue, Writing Creates Ecology and Ecology Creates Writing. *TEXT*, 20.

Somerville, M. (2013). *Water in a Dry Land: Place-Learning Through Art and Story*. London: Routledge.

Acknowledgements

My first acknowledgement must be to my river, the section of the Hawkesbury Nepean River system that is closest to my home in Emu Heights in Sydney's west, and to the rocks and stones that make up the bedrock of my river, its elemental healing place. There are many other rivers and waterways, plants, animals, birds, insects, children, and humans who have brought this river book into being. The Sydney Bass Fishers have been a central part of my riparian zone bushcare care learning. I thank Sue Collins who assisted with the original proposal, the co-authors who participated in the writing of chapters, Jacinta Tobin, Leanne Tobin and Susan Nordstrom. Others who have helped to bring this work into writing include Alison McConnell Imbriotis, whose patient reading of first draft writing never failed in its encouragement, no matter how difficult it seemed. Alison also helped with proof reading as did Susan Germein and Kerith Power and I am very grateful for that. Many others have offered me friendship throughout the challenges of writing that has enabled me to stay with the project, to all of these people I am grateful for their everyday loving kindness. And to the editors, Leila Walker and Rosie Anderson, and Susan Dunsmore whose encouragement and assistance have been so important to the final form of the river book.

1 A blue literature review

Introduction: a blue literature review

The blue literature review responds to the riverlands as a theoretical, material and embodied space to search for the literature of its becomings. It enters the space of the riverlands, finding a blue bower, and like a bower bird, collecting treasures, stealing others' words to lure the reader into the key concepts that frame the book: 'riverlands', 'Anthropocene', 'walking' and 'waterways as places of becoming'. In doing this, the blue literature review draws on 'the erotic energy of the unformed chaos of the world' by constructing a bower as the frame: The body is joined to the chaos of the earth through the pause enabled by the frame. The upright structures of the bower are the holding space for the ideas, written in a bower of words at the beginning and end of each of the key concepts. The paragraphs within the sections are the treasures upon which this performance rests, the words that provide the substance or ground of the performance. The headings are offered to highlight the blue treasures, to lure the reader into the ritual of the blue literature review and in between, the river adds its own literature review in its small river texts.

Collections of blue

Walking past morass
of tangled blackberry vines
blue chip packet blue plastic bag
form trail to carpet of deep blue
scrambling into prickly vines
deep blue carpet becomes
collection of blue
deep blue pegs more
and more deep blue pegs
blue bottle tops
scatters of blue plastic tape
and four perfectly constructed bowers

finest of fine twigs neatly inserted
into earth to form a holding place
to lure females into the frame.

The satin bowerbird's collection of blue (Figure 1.1) presented itself when walking the riverlands one day, struggling with how to write a literature review of the riverlands, desiring rather to presence the world in this writing.

The satin bowerbird

Searching the scientific literature, I discover a study of the satin bowerbird, (*Ptilonorhynchus violaceus*), and the relationship of their bower-making to the artful excess of natural selection. Satin bowerbirds live in eucalypt forests along the east coast of Australia. They are polygynous and during the mating season, male satin bowerbirds build stick structures to lure as many females as possible into the mating ritual. The bowers have a characteristic platform base and two parallel vertical walls forming a central avenue (Figure 1.2). Males adorn their bowers with various natural and man-made decorations and perform elaborate courtship displays on their bower platforms. Mating success is skewed among males, with only a few individuals obtaining most annual matings. Many aspects of the males' displays are correlated with mating success, including the type and number of bower decorations and the quality of the bower. Interactions among male satin bowerbirds are common, as males steal bower decorations from

Figure 1.1 Riverlands bower

Figure 1.2 The bower

one another and destroy the bowers of rival males (Wojcieszek et al., 2006, 175). In this blue (anti-) literature review I gather treasures like the bowerbird's blue objects, to lure the reader into the spectacle of becoming, stealing words from others, gathering them up as they appeal to me, taking pleasure in the assemblage of ideas, sensations, experiences, drawing on the chaos of the world.

The chaos of the earth

Drawing on Deleuze and Irigaray, Elizabeth Grosz theorises how 'living bodies draw from the chaos of the earth to extract from it something not so much useful as intensifying, a performance, a refrain, an organisation of colour or movement that induces art' (2008, 3). This roots art in the 'superfluousness of nature, in the capacity of the earth to render the sensory superabundant' (ibid., 17). Art draws on the erotic energy of the unformed chaos of the world by making a frame. In this act, the body is joined to the chaos of the earth through the pause enabled by the frame. The libidinous energy of chaos also serves to disrupt the frame as art breaks through systems of enclosure to enable something of the chaotic outside to reassert and restore itself in and through the body.

Riverlands

The bower

The riverlands at Emu Green on the Nepean River is neither fully water, nor fully land. It is the flood plains of the Nepean River that cannot be overwritten with the grid of human development. In this contested space of an urban commons, human and more-than-human come into relation, mediated always by the inevitable flow of the river, always already there, long before us and long after we are gone. Water, animals, trees, all the river's creatures and its humans produce each other within the material bodies of things and elements that come together there. It is the space of a borderlands where difference is endlessly produced as beings and elements merge and emerge in order to produce the new.

Fluffy cream brushes bloom
on sacred river gum spirit trees
wattle bird squawks
birds flitter tree to tree
overhead in
cacophony of bird song
blue heron lifts fine legs
spreads wide grey wings flies
off soundlessly into nearby trees
at dawning of new day
in the river's lands.

Ecotones

The word 'ecotone' originates from the Greek roots '*oikos*' (home) and '*tonus*' (tension). Definitions include boundary regions, borders, meeting zones, transitional zones, tension zones, zones of intermingling and zones of transgression. They are considered to be dynamic zones of interaction between communities, which are unstable over time (Kark, 2013). Ecologists argue that boundaries are ubiquitous and essential in nature, in particular, because of how they differentiate habitats, regulate the transfer of material, organisms or information, and affect neighbouring systems. Boundaries between natural ecosystems form in response to topographical, hydrological, geological or climatic variation in a landscape. Scholars study these transitional and dynamic boundaries as ecotones. The emergent nature of these zones supports and features the inhabitants and properties of the overlapping systems, along with their own unique array of characteristics, including species (Simmonds and Gazley, 2018).

Flock of peewits and
gaggle of tiny black birds
potter on grasslands

wild yams bloom white and pink on
long feathery stems
wild fennels aniseed smell perfumes
ducks camouflaged in brown green pond
dabble and flicker as
blue black crows fly gracefully by and
caw caw echoes across the valley.

Edge effects

Ecotones do not merely represent a boundary or an edge, rather an active interaction between two or more ecosystems which create unique areas emerging from the force of the interactions among the adjacent ecological systems. Due to these characteristics, ecotones frequently support comparatively large amounts of diversity, activity and energy, giving rise to, and then reproducing, the emergence of the unique landscape features and species which come to inhabit these areas, a phenomenon known as *edge effects*. In ecological terms, edges are zones of transition from one ecosystem to another, areas where two different types of habitat, or successional stages, meet and intergrade. A well-known characteristic of ecosystems is that these edges often exhibit high levels of productivity and species richness or biodiversity (Turner et al., 2003).

Dry blond of late winter grasses
lit by vivid pink of vetch
bright gold of coreopsis
yellow fireweed daisies
sprinkled among
vivid pink vetch
scatter of blue stars
and clusters of brilliant purple
tall low here there and everywhere
in the river's lands.

Speciation and global change

Ecotones are 'natural laboratories' for studying a range of evolutionary processes, such as the process by which new species form, also termed speciation. This process is of major interest to evolutionary biologists who define three major types of speciation: allopatric, parapatric, and sympatric. These models are based on the degree of geographical subdivision between populations that lead to the formation of new species. Substantial conservation attention has been given in recent years to the understanding and mapping of biodiversity patterns and the underlying processes, and towards predicting the effects of global change. Ecotone and boundary regions, where change, shifts, and variability occur naturally in both space and time, could serve as useful models for understanding, monitoring and predicting the response of individuals, populations and communities to changing environments (Kark, 2013).

Purple top verbenacae
and purple verbena rigida
tiny clusters of pink star flower
dark pink of wild raspberries fruiting
orange-spotted butterfly
hovering over flowers
blue wrens flitting on tree stumps
bright pink of fluffy scotch thistle
flower on prickliest of bushes
beside soft feathery sweet smelling
wild fennel
in the river's lands.

Riparian zones as ecotone

Riparian zones are the interface between aquatic and terrestrial environments. The riparian zone encompasses the stream channel between the low and high water marks and that portion of the terrestrial landscape from the high water mark towards the uplands where vegetation may be influenced by elevated water tables or flooding and by the ability of the soils to hold water (Weissteiner et al., 2016). They are transitional areas occurring along land and freshwater ecosystems, characterised by unique soil, hydrology and biotic conditions strongly influenced by the stream water. These areas are unique and diverse, and are often the most fertile parts of the landscape. They exert important influences on the waterways they adjoin by mediating the bi-directional flow of matter and energy between the water body and the surrounding hinterland. Riparian research demonstrates that intact riparian zones are critical to aquatic-terrestrial ecosystem function and ultimately, to waterway health.

Tease our competing grasses
from dry ground
around thirsty roots of
coffee bush Australian indigo
cheese tree forest red gum
manna gum
and newly opening pale red flowers
of bottlebrush
until all grasses are removed
and prickly biden pilosa calls from
acacia floribunda, ti trees, and callistemon.
Tender tended weeding in conviviality
at the river's edge.

Material flows

The riparian zone is intrinsically linked to the surrounding catchment through material flows (water, soil, nutrients, chemicals), an extensive ecotone, and animal movement. Compared to the surrounding landscape, riparian zones can be relatively nutrient-rich environments and hence higher quality habitat. Soil and nutrients are transported from surrounding areas and deposited in riparian zones. These fertile alluvial soils and often moister conditions result in productive environments for both plants and animals. Variable flood duration and frequency, and concomitant changes in water-table depth and plant succession, create an environment that is a complex of shifting habitats, established and destroyed on different spatial and temporal scales (Naiman and Décamps, 1997).

> Water's flow on river stones
> sparkles silver in early morning's
> play of light
> as slow current carries
> flotsam and jetsam of
> travelling water weed
> salvinia and water hyacinth,
> bubble florets,
> fallen leaves and flower petals
> flow slowly by
> as red bottle brush comes into bloom
> flies buzz and birds chatter
> at the river's edge.

Movement of plants and animals

Riparian lands play a crucial role as a corridor for the movement of plants and animals. They wind their way through diverse ecosystems forming natural corridors linking habitats and providing routes for dispersal or migration of terrestrial and aquatic animals. Vegetation in the ground and shrub layers, rocks and fallen timber all provide habitat for invertebrates, frogs, lizards, snakes and small ground-dwelling mammals. Some water-birds such as herons, rails, bitterns and warblers use thick reed beds and other vegetation fringing watercourses for shelter, foraging and nesting.

> Ducks skitter
> butterflies flutter
> whip bird calls
> water gurgles
> finches twitter
> dragonflies hover
> water weeds float

lizard scampers
mulberries ripen
bees buzz
fishes wriggle
at the river's edge.

Most productive eco-systems on Earth

Some animals rely on riparian lands for their entire lifetime whereas others may only need them at particular times of day, certain seasons or life stages. Moisture, including soil moisture, rich organic matter, soils and nutrients, is replenished by periodic flooding. Shading capacity and moisture reduce extremes of temperature and humidity, providing more balanced conditions for plants and animals. As a result of greater availability of water and soils rich in nutrients, riparian lands are among the most productive eco-systems on Earth.

Mist flows back and forth
rising from warm water's still surface
as sun glows golden behind
dark outline of trees and
tiny grebes appear and disappear
lost in the mists of time as
whip birds call and respond
call and respond call and respond
over and over again while
tiny mid-river islands come alive
with twittering of tiny birds
in sunrise at the river's edge.

Habitat

Shallow lentic water with vegetation provides habitat for frogs and their tadpoles, and for aquatic macroinvertebrates. Outcropping rocks and river pebbles (cobbles) provide important habitat for reptiles. Vegetation with a complex structure (herb, shrub and tree layers), which often occurs in the riparian zone, is important nesting and foraging habitat for many bird species. Large trees in or beside watercourses provide nest sites for raptors or nesting hollows for possums, bats and birds, such as cockatoos, kingfishers and some duck species. Structural complexity is created by the presence of trees of different ages: tree hollows, standing dead trees, a patchy shrub layer, a species-rich understorey of grasses and forbs (herbaceous flowering plants), fallen timber and water, all of which provide shelter, food or nesting sites for animals (ACT Government, 2002).

Moor hen's call echoes across
clumps of dry dry grasses

river is green and cool
light breeze brings river's breath.
Cormorants moor hens ducks
in their spots doing
their water bird things
in quiet rustle of water on rock
flapping of ducks upstream
chorus of moor hens
playing playing high-pitched
river's music
as squealing black cockatoos
land in sheoaks
at the river's edge.

Multispecies and interdependence

Animals are essential for pollination and dispersal of many plants and are involved in nutrient recycling and maintenance of soil condition and water quality. While some animals are largely restricted to the riparian zone, other animals living in the surrounding landscape use the riparian zone and associated rivers or streams as a source of food, water, shelter or nesting sites. Other animals move mainly between the water and adjacent riparian areas.

Insects, other macro-invertebrates and microbiota are vital for healthy ecosystem function. They are essential for pollination and reproduction of many woodland plants, are involved in nutrient recycling through the breakdown of dead plant and animal material and are the main food of many aquatic and terrestrial animals. Terrestrial invertebrates are more diverse and abundant in habitats with mature trees (loose bark) and a well-developed ground cover of leaf litter, rocks, logs, branches or tussock grasses. Fallen timber, leaf litter and other ground cover provide important habitat for a range of invertebrates, many of which depend on dead wood and leaves for their survival. These invertebrates, together with microbial organisms and fungi, are important in the breakdown of timber (Araya, 1993) and recycling of nutrients back into the soil. In turn, invertebrates provide the main food source for a range of birds, reptiles, amphibians and some mammals.

Hot late spring afternoon
amongst boulders
under cool deep shade
of white cedars perfume
wide-spreading branches
native frangipani's clusters of golden
flowers dripping with nectar
as birds and bees are
called by its fragrance to
eternal cycle of pollination

and reproduction.
Tumble through boulders to
dangle feet in cool water
at the river's edge.

Human enmeshments

Riparian areas are vulnerable and easily degraded. The physical connectivity and narrow width of riparian zones leave little buffer against impacts occurring in adjacent ecosystems. Expanding urban development increases pressure on catchments, aquatic and wetland habitats and remnants of native vegetation, whereas construction of roads and other urban infrastructure leads to increased fragmentation of habitat. Fragmentation of riparian zone habitat reduces the value of these zones as corridors connecting different areas and habitats. Loss of habitat diversity in the riparian zone has a major impact on riparian fauna. Ecosystems with a complex 'architecture' support more species than ecosystems that have been simplified, and many species require a complex vegetation structure to meet their habitat requirements.

Five plastic bags
one red cardboard Jatz packet
one empty silver foil wrapper
one silver-lined Kellogg's
Nutrigrain wrapper
one large plastic bottle
nine pink plastic cups
one brown glass bottle
one yellow and silver drink can
and one small corn can
gather at the river's edge.

Small lowland wetlands have been drained and/or invaded by weeds. Creeks in urban areas have been straightened, piped or converted into concrete-lined drains. Large streams and rivers have flows altered by dams, temperatures lowered by cold-water pollution or raised following the clearing of riparian vegetation, and water quality is affected by sedimentation due to erosion or catchment degradation. Damage from urban development includes invasion by pests and feral animals such as rabbits, weeds such as privet, or overuse by recreational activities. Foxes, cats and dogs are known to prey on a range of native animals in riparian habitats including mostly ground-dwelling small mammals, reptiles, frogs and birds commonly found on the ground or in lower understoreys, and occasionally bats and small arboreal mammals.

Concrete drainway
creates its own installations
as shopping trolleys

gather floating twigs, wire, polystyrene
becoming enmeshed
in eternal flows
of bottles, cans, building materials
chicken wire, children's toys
and new red shopping trolley
appears in Lapstone creeks'
remnant flow.

Transcorporeal bodies

Potent ethical and political possibilities emerge from the literal contact zone between human corporeality and more than-human nature. Imagining human corporeality as transcorporeality in which the human world is always emmeshed with the more-than-human world underlines the extent to which the substance of the human is inseparable from 'the environment' as a world of fleshy beings with their own needs, claims and actions. Transcorporeality reveals the interchanges and interconnections between various bodily natures, and opens up a mobile space that acknowledges the often unpredictable and unwanted actions of human bodies, non-human creatures, ecological systems, chemical agents and other actors (Alaimo, 2017, 2).

The bower

The literature of the riverlands states, over and over again, the idea that ecotones and riparian zones in particular are place of extraordinary species diversity, and even the becoming of new species. The riverland's riparian zone is described as a complex of edge places, where difference comes together in sometimes frictional ways. Difference in such edge places is a source of rich abundance, mobility and change where species interact and adapt, and new species are brought into being. The transposition of humans into these fragile and unique ecosystems positions the human as part of the larger trans-corporeal body of a whole ecotone. It is possible that humans might also become different in these edge places, transformed within the rich abundance of speciation, a new human species for the Anthropocene, the age of human entanglement in the fate of the planet. A human species with new ways of being, knowing and doing, ways that participate in the rich abundance that the endless proliferation of difference in the life of these places creates.

The Anthropocene

The bower

Described as 'a new phase in the history of both humankind and of the Earth, when natural forces and human forces became intertwined, so that the fate of one determines the fate of the other' (Zalasiewicz, Williams, Steffen and Crutzen, 2010, 2231), the Anthropocene is proposed as a new geological age. In the face of the enormity of planetary crises, the focus on the Anthropocene can be a tool for innovation and imagination to generate emergent constellations of life and knowledge because climate change is not only a change of the climate but a change in the very ways in which we think, requiring us to develop new concepts of the human, new figures of life and new understandings of what counts as thinking (Colebrook, 2010).

The mattering of time

Formal recognition of the new epoch of the Anthropocene depends on establishing a date for which there is geological evidence of time's mattering and it is this intimate relationship between time and matter that is the first disruption to thought. This mattering is the not-yet of time, what will be visible in layers of the Earth's geology some time into the future, maybe even after the end of the human species.

There are many alternative suggestions about how the matter of time might mark the onset of the Anthropocene. It might be visible in plutonium fallout from 1950s bomb tests in marine or lake sediments, ice layers or stalagmites and stalactites; in layers of remnant plastics, or in some kind of carbon signature marking the rapid rise in CO_2 emissions (Amos, 2016).

Plastic aluminium and fly ash particles; fossils; artificial radionucleides; and changed carbon and nitrogen isotope patterns are signals in recent strata of changes made to the Earth system by human activity. Most of these signals will remain in the Earth's strata as a permanent record of past processes (Schneiderman, 2017).

Waste constitutes perhaps the most abundant and enduring trace of the human for epochs to come, whether in the form of mining, nuclear, industrial, hazardous, sewage or municipal, and whether it is dumped, landfilled, incinerated or buried deep underground. The real provocation of the Anthropocene is not that we are leaving a message for some imagined future humanity to decipher but that we are bequeathing a particular futurity (Hird and Zahara, 2017).

The mattering of life and death

The major boundaries between geological eras are marked by mass extinctions – times when the Earth loses more than 75 per cent of its species in a geologically short interval. The Anthropocene is commonly regarded as the epoch of the sixth mass extinction because the world's species are vanishing at an unnaturally rapid rate. However, the most devastating mass extinction marks the end of the Palaeozoic era (approximately 250 million years ago). Geologists often refer to this extinction as the 'Great Dying', because 96 per cent of marine species died out in less than 200,000 years. It took millions of years to recover from this extinction event, and all life on Earth today is descended from the 4 per cent of species that survived it (Schneiderman, 2017).

It is the fossil record that inscribes these mass extinctions into readable form in the Earth's sedimentations. Foucault's fossils fundamentally disrupt our taken-for-granted assumptions about life and death. As the biological record of catastrophic extinction, Foucault's fossils become the strange time-twisting mirrors of the 'ends of man'. This 'lithic conviviality' (man reading fossil) can be understood as a mad palaeontology, a speech after death, the proleptic trace of our own extinction. Like a ghost, it marks what Foucault calls that uncertain frontier region where one does not know whether one ought to speak of life or not. The fossil lingers as a strange remnant of something we call life within a frame where that something can no longer be thought (Huffer, 2017).

The mattering of the human subject

Anthropocene thinking shifts understanding of the human, who occupies only a microsecond of geologic time, poised at this moment between the geological past and an uncertain future (Schneiderman, 2017). To live is to tend towards *indifference*, where tendencies and forces result less in distinct kinds than in complicated, confused and disordered partial bodies. One might think here of the human body whose vital attachment to the Earth, food, sex, technology and other humans not only disperses the self across a series of connections but also operates as much to diminish as to intensify stability (Braidotti, 2017).

The immersed human subject of the Anthropocene can be seen as an intermingling of biological, chemical, and climactic processes, which are neither simply natural, nor managed by human intention. The transcorporeal subject emerges from environmental health and environmental justice movements, including the citizen-scientists who must discern, track and negotiate the unruly substances that move across bodies and places. Thinking the subject as a material being subject to the agencies of the compromised, entangled world enacts an environmental posthumanism, insisting that we, as bodies and minds, are inextricably interlinked with circulating substances, materialities and forces (Alaimo, 2017).

The relational capacity of the posthuman subject is not confined within our species, but it includes all non-anthropocentric elements, starting with the air we breathe. Living matter, including human flesh, is intelligent and self-organising, but it is so precisely because it is not disconnected from the rest of organic life and connects to the animal and to the Earth. We think with the entire body, or rather, we have to acknowledge the embodiment of the brain and the embrainment of the body. We may yet overcome anthropocentrism by becoming anthropomorphic bodies without organs that are still finding out what they are capable of becoming (Braidotti, 2017).

The bower

The proposed geological age of the Anthropocene disrupts fundamental taken-for-granted assumptions about what it means to be, and to know the world, including concepts of time, life and death, and the human subject. It has acted as a substantial force producing collective action, and both new and renewed forms of thinking in western posthuman, more-than-human, and new materialist theorising, common worlds, and global indigenous understandings which have always understood the human as an inseparable part of the planet Earth.

Walking (and stopping)

The bower

The only way to think and become differently in the face of the enormity of anthropogenic changes to the Earth's climate, and the consequent loss of biodiversity, is through becoming embedded and embodied in the intimacy of local places. Ideas, concepts and theories emerge in the enactment of walking every day in the riverlands close to my home. Walking produces embodied thought, and body becomes other to itself with trees, plants, water, rock and stone, insects, lizards, snakes, smells, sound, seasons, quality of air, and so on, always forever changing, each day becoming something different.

Walking in the Anthropocene

At first glance, 'walking' and the 'Anthropocene' do not seem to go easily together. After all, walking is slow, fragile, unreliable – it's hardly a mode of movement suited to grappling with the pressing issues the Anthropocene heralds. Surely we don't have time to dawdle in the face of the urgent politics of global change? But in another sense, walking might be exactly what we need. The slow, engaged and engaging attributes of walking might indeed help to enhance our connectedness with the world in embodied and creative ways. The mode of walking and wayfinding appropriate to the Anthropocene isn't a headlong rush to get to somewhere 'better' or the conceit of thinking that we have the answers. Rather it's a studied movement of the here and now, a fragmentary, wandering, lively, embodied and relational process, a respectful movement that puts emphasis on sensory, contingent and fragile encounters conjured through making our way, alongside others through time and space, here and now (Instone, 2015).

Dialogue with loss

My forays are motivated by a sense of grief and the loss of local and global ecosystems and the escalating rate of species extinctions. The black birds that fly overhead as I walk – the Carnaby's cockatoo and the forest red-tailed cockatoo, are two such species that, due to land clearing and climate change, are predicted to become extinct within the next 50 years. The walking project is an attempt to find a horizon within this dialogue of loss, and within the altered geography of local and global environments (Chinna, 2015, 11).

Walking as methodology

Walking methodologies are epistemological, ontological, ethico-political orientations to how one engages with research. The concept of the more-than-human emerges at a time in scholarly debates that seeks to challenge and de-centre human exceptionalism, taxonomies of intelligence and animacy, and the distinctions made between humans and non-humans, nature and culture. Rather than view the inhuman as the opposite of the human, the inhuman becomes a process by which human and non-human frictionally come together. The inhuman is a process of thinking differently where intimacy and difference come together (Springgay and Truman, 2018, 5).

Friction is a force that acts in the opposite direction to movement. Friction exists every time bodies come into contact with each other, like different strata grinding against one another (ibid., 8–9).

Ethical walking

In thinking about how to move collectively, of being accountable in the presence of others, walking enacts situated and contingent ontologies between land, peoples, and nonhuman others. Walking-with demands that we forgo universal claims about how humans and non-humans experience walking and consider more-than-human ethics and politics of the material intra-actions of walking research. You could walk-with alone (ibid.).

Walking with multiple others

Whether we realise it or not, whenever we walk, we are walking alongside multiple others, human and non-human, and how we move is likewise not only a human achievement but shaped by the more-than-human worlds through which we step. Walking-with others involves our hearts as well as heads, legs, guts and minds. It combines rational and more-than-rational knowing (Instone, 2015, 137).

Walking as embodiment of time and space

Walking slows you down, time passes differently, and mind and body are merged in the effort to cover ground and take in the surroundings. That is, every step embodies time as well as space, each step meshing things past and those to come in an ongoing process, each step participating in the making of worlds and in the process, knitting together responsibility for past, present and future (ibid., 135).

Walking as moving body-without-organs

Walking: putting one foot in front of the other, poising the weight on one foot and then tilting the body forward with the other foot, swinging this foot in front of the body and placing it on the ground to prevent falling. Walking reconstructs Galileo's pendulum, the legs move through time and space, marking the movement over grass, stones and hills, and through wind which is air moving through time and space. I walk slowly and time dissipates to the stillness of my breath wrapping around me in tight coils. As I pick up speed, time gathers to meet me, rushing round the curvature of the Earth (Chinna, 2015, 8).

Stopping to deep listen

Walking also involves sitting in places and engaging in deep listening, exploring the layers of sound from the loudest to the barely perceptible. It involves tasting plants, water and air, touching these same; noticing wind currents, movements of birds and people, and all the colliding and conflicting sensations that the body experiences in an environment. I record my observations in a series of notebooks and use these notes to later inform the poems I write (ibid., 9–10).

Traversing terrains of world and mind

For the creative practitioner, walking reintroduces the body as a fundamental definer of experience. The walker uses the body as a divining rod, pacing through time, noticing what needs to be noticed. Exploring the world is one of the best ways of exploring the mind, and walking traverses both terrains (ibid., 9–10).

Poepatetics

To define my walking/writing practice I constructed the neologism poiepatetics or the poetry of walking: poie from poiesis, i.e., making, and patetics from peripatetics or walking, travelling, a person who walks or travels about. More simply put, poiepatetics is a making from walking. The production of poems is closely related to the walking practice, it is the physical experience of the places within which I walk that really provides the basis for most of the pieces (ibid., 8–9).

Writing on water

First you must wade through the minutiae
copepods, water boatman, and backswimmers
You may be bitten by fleas
reborn after aestivating for two hundred years.
Remember frogs cannot swallow
with their eyes open so they may not see you coming.
Sift out the sediment (this may be achieved by taking off
your shirt and straining the water through the cloth.)
There is a lot to know before you can start
water can kill as easily as quench.
water can be very old

water makes ink run
will dissolve paper.
One letter too many or too few
can change the whole meaning.
Until it dries out you may not be able to understand
what water has to say.

(ibid., 15)

The bower

Walking as a fragmentary, wandering, lively, embodied and relational process, as a dialogue with loss, and with more-than-human worlds is integral to these authors' approaches to the global challenges of the Anthropocene. Walking offers alternative methodologies, ways of being in, and knowing, the world. Ethical walking means walking with others, human and non-human, and welcoming the frictional encounters inevitable when different bodies come together. Walking is a way of moving in time and space, through which the body becomes other to itself, a body without organs, connected to ground, stones, air, wind, bird song, weather, grasses, light and shade, time of day. Walking can uncover hidden histories of waterways, often submerged beneath concrete, turned into pipes and drains, through stopping for deep listening. Walking and stopping for deep listening produce encounters of becoming.

Waterways as places of becoming

The bower

Deleuze and Guattari's 'becoming' offers an overarching theoretical/conceptual approach that enables the integration of new materialism, common worlds and Aboriginal ecophilosophies described above. The complex moment-by-moment intra-actions of all of the elements, living creatures, plants and animals, atmosphere, air, light, and humans, that come together frictionally in the place of the riverlands can be embraced in Deleuze's becoming. As with the bowerbird's ritual performance, the intention in exploring becoming is to entice, to enact, rather than explain, to offer quotes that animate as forms of poetry. Each Deleuzean quote is linked to momentary encounters from my river walks produced as river's texts, an entry into fluid thinking that flows and eddies like water.

Encounters of becoming

We can be thrown into a becoming by anything at all, by the most unexpected, most insignificant of things. If we imagined the position of a fascinated Self, it was because the multiplicity towards which it leans, stretching to breaking point, is the continuation of another multiplicity that works it and strains it from the inside. In fact, the self is only a threshold, a door, a becoming between two multiplicities. Each multiplicity is symbiotic; its becoming ties together animals, plants, microorganisms, mad particles, a whole galaxy (Deleuze and Guattari, 1987).

Riverlands, 18 January 2019, 7.40 a.m.

No one is here but
caw caw lonely call of crow flies overhead
wild ducks cruise by
dip and splash dip and splash
silver fish leaps up and
disappears under water
dome of turtle shell on bank
webbed feet poking out
reptilian head glinting eye
curls back inside shell with
strange stillness at my presence
crackly cicada shell caught on socks
as summer song of cicadas rises and falls.
There is never no one here.

Becoming-animal becoming imperceptible

Becomings-animal are basically of another power, since their reality resides not in an animal one imitates or to which one corresponds in themselves, but in that which suddenly sweeps us up and makes us become – a proximity, an indiscernibility that extracts a shared element from the animal far more effectively than any domestication, utilisation, or imitation could: If becoming-woman is the first quantum, or molecular segment, with the becomings-animal that link up with it coming next, what are they all rushing towards? Without a doubt, towards becoming-imperceptible. The imperceptible is the immanent end of becoming, its cosmic formula (ibid., 279).

Riverlands, 1 March 2017, 7.40 a.m.
Beady eye catches mine
of reptilian body draped
long to cleft of trunk
clawed feet clutching rough
crinkly bark of casuarina
ghoulish smiling face turns
webbed with jagged dinosaur crests
becoming molecular imperceptible
to each other until one final move

puffs out reptilian throat and drops
into water as we become
once again two separate bodies
from different worlds.

Spatio-temporal assemblages of events as encounters

Climate, wind, season, hour are not of another nature than the things, animals, or people that populate them, follow them, sleep and awaken within them. This should be read without a pause: the animal-stalks-at-five-o'clock. The becoming-evening, becoming-night of an animal, blood nuptials. Five o'clock is this animal! This animal is this place! (ibid., 263).

Riverlands, 27 July 2018, 7.56 a.m.
Mist rises as cold air
touches warm surface
at dawn of new day
water catches sunlight in silver ripples
and elegant long-necked white egret's
perfect reflection on still surface
rumble of trucks blows on light breeze
egret flies off to land close by
walks on long fine legs to river bank
becomes hidden in bushes
until I get up to leave and
egret flies off from hidden place in
a unison of leaving.

Time of pure event or becoming

In the first place, the verb in the infinitive is in no way indeterminate with respect to time; it expresses the floating, non-pulsed time proper to Aeon, in other words, the time of the pure event or of becoming, which articulates relative speeds and slownesses independently of the chronometric or chronological values that time assumes in the other modes (ibid., 263).

Riverlands, 27 July 2018, 3.59 p.m.

Sea eagle soars down river highway
light breeze blows
dank river breath into mine
so calm quiet and warm this afternoon
sitting at water's edge
too cold to swim
will get into water soon with

change of season but now it's only
swimming in swirling currents
of cold white frosty air
alternating with
flow of warm currents
at this edge of a new season.

Memories of a molecule

Becoming-animal is only one becoming among others. A kind of order or apparent progression can be established for the segments of becoming in which we find ourselves; becoming-woman, becoming-child; becoming-animal, -vegetable, or -mineral; becomings-molecular of all kinds, becomings-particles. Fibres lead us from one to the other, transform one into the other as they pass through doors and across thresholds. Singing or composing, painting, writing have no other aim: to unleash these becomings (ibid., 272).

Writing as becoming

The fish is like the Chinese poet: not imitative or structural, but cosmic. They retain, extract only the essential lines and movements of nature; they proceed only by continued or superposed 'traits' or strokes. It is in this sense that becoming-everybody/everything, making the world a becoming, is to world, to make a world or worlds, in other words, to find one's proximities and zones of indiscernibility. To reduce oneself to an abstract line, a trait, in order to find one's zone of indiscernibility. One is then like grass: one has made the world, everybody/everything, into a becoming, because one has made a necessarily communicating world, because one has suppressed in oneself everything that prevents us from slipping between things and growing in the midst of things. Eliminate everything that exceeds the moment, but put in everything that it includes – and the moment. To be present at the dawn of the world (ibid., 280).

Blocs of becoming

The chapters of *Riverlands of the Anthropocene* are conceived as blocs of becoming, a passing through, not an arrival at a fixed point, composed of interdependent symbiotic relations, liberating, releasing the individual, autonomous beings from their isolation. What is real is the becoming itself, the bloc of becoming, not the supposedly fixed terms through which that which becomes passes (Deleuze, 1987). The chapters take up the motifs of the blue literature review, responding to the waterways of our rivers as a global space, and to the intimate local of Emu Green riverlands, as theoretical, material and embodied spaces in their moment-by-moment becomings.

The structure of the book

Chapter 2, 'Rivers of the Anthropocene', flows from the intimate local of the Hawkesbury Nepean riverlands at Emu Green into the interconnected matrix of rivers and oceans of our blue planet to ask, how are the major iconic rivers of the world becoming in the age of the Anthropocene? I do this digitally, lying on the red sofa in my house that overlooks the riverlands, flowing with the *Rivers of Life* television series. Watching episode after episode, over and over again, I am enmeshed in their becomings as the rivers' images, sounds and stories wash over me until eventually I write handwritten skeletal notes. These notes function as mobile images forming into words that are transformed into each iconic river's own *river text*. Further searches into the digital world of Google satisfy my desire to enter even more into becoming with the world's major rivers. Each river invites its own distinctive embodied knowing and poses questions that flow throughout the chapters of the book.

Chapter 3, 'Watery beginnings', poses the questions asked in consort by all of the rivers of the world, what are water's stories?, what does water do?, and how does water shape the human in its eternal flows through the bodies of the world? In response to the rivers' questions, I travel with the flow of the Hawkesbury Nepean River, from the freshwater of my daily swim hole at Emu Green until it reaches the sea at Broken Bay, where freshwater and saltwater meet. Here in this edge place, new stories of water emerge: from water's beginnings of life on Earth; the beginnings of human life in amniotic fluid; the waters of first contact; the waters that carried my ancestors from Scotland; and the watery beginning of the mega city of Sydney in Warragamba Dam. Waters' stories traverse two very different time zones, held in the space between the geological beginnings of life, and the formation of the mega city as it exists today.

Chapter 4, 'Songlines', invites the reader into a becoming through walking in deep time, and through walking the contemporary songlines of the Darug peoples today, in answer to the challenge of the mighty Nile River, that we do not need another explanation for what we have lost, but an exploration of what we may find. Songlines are walking trails that follow the stories of the great ancestral beings as they created the world and all that is in it. The songlines that once echoed throughout the river's valley are now gone. Songlines, however, are resilient, they live on, they are recreated in song, story, art and language. New songlines hold the contemporary history and cultural memories of singing the rivers in song and ceremony for Darug peoples today. Songlines invite the reader to meditate on the presences and absences of these river places with Darug artists, song writers and cultural knowledge holders.

Chapter 5, 'Riverlands' watery ways', ask the Hawkesbury Nepean River's own question, how to bring the river to you, its reader. It is a methodological becoming within the intimacy of the Hawkesbury Nepean River's watery ways at Emu Green. It explores how the river creates its places on the songline where water and land meet and everything is in a moment-by-moment relation with water, rock and stone, soil, weather, air and light, seasons, bird song, cicada call, plants, trees, litter, in its ever-changing becomings. A methodology of 'spots' is developed to engage with emergent representations of the sensory corporeal experience of the riverlands' watery ways. The proliferation of still photos, small videos, and skeletal river notes are filtered through seven spots: *Bedrock; New swim hole; Regeneration; River's crossings; Litter; Drainway*; and *Spot of death*.

Chapter 6, 'Bedrock's sacramental becomings', is a wondering about how our rivers become sacred and what are the elements of their sacramental becomings, in response to the Ganges' question, how do our rivers become sacred? It is the story of a long slow time of illness and healing when I reached rock bottom, the bedrock of my being. Every day at dawn and at midday I was compelled to go to the closest place on the river, marked by its dense row of casuarinas. In the warm seasons I slide in and immerse my body, paddling up and down, up and down, and when the water becomes too cold, I sit close to the river and become lost in the reverie of its slow flow. During this long slow time of healing, this place on the river became *Bedrock*, the space of a river's sacred becomings, recorded in river texts, still images and short videos.

Chapter 7, 'The river's crossings', invites the reader into a playful multi-species, multi-elemental becoming that uses the metaphor of the site of the river crossing as a means to 'cross to the other side' of being and knowing. Children become sand birds, rock jagwars, mud girls, seal babies, see a crocodile, and a mermaid, crossing to the other side of the human species with relative ease. Animals insert themselves with equal ease into these stories of human child becomings. The river's crossings echo the Yangtze River's question about the salvation of our rivers through the stories of human becoming animal and animal becoming human. It is experimental, playful, and transgressive in its writing, challenging the writer and reader to think themselves into becoming with the life of the other.

Chapter 8, 'Global materialities: the artful excess of river's litter' dwells in the realm of litter's inanimate manifestations that tell their own stories of movement and flow, stories of the rivers. They are stories of the inhuman within the human. Litter inscribes its own literacies, some litter with printed text of its own, others with brand names that tell of their multinational affiliations and global flows. Litter's textual materialities call up an exchange with Susan Nordstrom from the Mississippi River in Memphis, Tennessee. Our rivers' waste meets symbolically on remote Henderson Island in the Pacific Ocean, the place with more plastic than any other surface on Earth. Plastic and waste call us back to our rivers, the Nepean and Mississippi, to think with waste. Our thought experiment with waste materialises transformative becomings that generate past-present-future affective residues of wonder about the materialities of litter and rivers. We create with bodies of waste, bodies of rivers, bodies of humans, and bodies of theory.

Chapter 9, 'Regeneration: of trees, weeds and tender intimacies', is set in the context of anthropogenic changes to the Earth's climate, and the regenerative possibilities for the planet of the planting and care of trees and other plants. The chapter explores these possibilities in the very local site of bushcare regeneration of the riparian zone of Emu Green by the Sydney Bass Fishers. Because of their far-reaching involvement in rivers and waterways, the Sydney Bass Fishers are unique in their approach to bush regeneration. For them, it is not only about the trees and weeds, and all the plants in between, but about the relationship between the waterways and the riparian zone. The chapter explores the multiple entangled relations between the plants, humans, the waterways and its creatures. The iconic River Rhine flows into and through this chapter in its question about how our rivers can be transformed through the care of their riparian zones.

Chapter 10, 'Life and death in the Anthropocene', is located in the great paradox of the Anthropocene, hovering in the inseparable intertwinement of the mighty Amazon River with its rainforest, and its global regulation of atmospheric cycles. It is the time of the sixth great extinction, 'a new phase in the history of both humankind and of the Earth, when natural forces and human forces became intertwined, so that the fate of one determines the fate of the other' (Zalasiewicz et al., 2010, 2231). As the greatest anthropogenic threat to the commons the world has ever known, it is also the time when the human is blessed with one last chance to become part of the planet. The chapter, 'Life and death in the Anthropocene' gathers up stories that have produced themselves as intact stories waiting to be told for the ending of the river's book. They are all stories that sit in the paradoxical space of the Anthropocene, where life and death intersect, where death produces life and life produces death. They are stories of *The Spot of death; The Flowers of eternal life; Burying gold; Mississippi floods*; and *the Gundungurra creation story*.

References

ACT Government (2002). ACT aquatic species and riparian zone conservation strategy. Available at: www.environment.act.gov.au/__data/assets/pdf_file/0005/576320/Action_Plan_29_Chapter_3.pdfAlaimo, S. (2017). Your shell on acid: material immersion, Anthropocene dissolves. In R. Grusin (ed.), *Anthropocene Feminism*, Minneapolis, MN: University of Minnesota Press, pp. 89–121.

Amos, J. (2016). Geologists search for Anthropocene 'golden spike'. *BBC News, Science and Environment*. Available at: www.bbc.com/news/science-environment-37200489

Araya, K. (1993). Relationships between decay types of dead wood and occurrence of lucanid beetles (Coleoptera: Lucanidae). *Applied Entomological Zoology*, 28, 27–33.

Barad, K. (2007). *Meeting the Universe Halfway*. Durham, NC: Duke University Press.

Braidotti, R. (2017). Four theses on posthuman feminism. In R. Grusin (ed.), *Anthropocene Feminism*. Minneapolis, MN: University of Minnesota Press, pp. 21–48.

Chinna, N. (2015). *Swamp: Walking the Wetlands of the Swan Coastal Plain*. Fremantle, WA: Fremantle Press.

Colebrook, C. (ed.) (2010). *Extinction: Framing the End of the Species*. London: Open Humanities Press.

Deleuze, G. and Guattari, F. (1987). *A Thousand Plateaus: Capitalism and Schizophrenia*, trans. B. Massumi. Minneapolis, MN: University of Minnesota Press.

Grosz, E. (2008). *Chaos, Territory, Art: Deleuze and the Framing of the Earth.* New York: Columbia University Press.

Hansen, B., Reich, R., Lake, P. and Cavagnaro, T. (2010). Minimum width requirements for riparian zones to protect flowing waters and to conserve biodiversity: a review and recommendations. Melbourne: Department of Sustainability and Environment.

Hird, M. and Zahara, A. (2017). The Arctic wastes. In R. Grusin (ed.), *Anthropocene Feminism.* Minneapolis, MN: University of Minnesota Press, pp. 121–146.

Huffer, L. (2017). Foucault's fossils: life itself and the return to nature in feminist philosophies. In R. Grusin (ed.), *Anthropocene Feminism.* Minneapolis, MN: University of Minnesota Press, pp. 65–88.

Instone (2015). Walking as respectful wayfinding. In K. Gibson, D.B. Rose and R. Fincher (eds), *Manifesto for Living in the Anthropocene.* New York: Punctum Books. pp. 117–122.

Kark, S. (2013). Ecotones and ecological gradients. In R.A. Meyers (ed.), *Encyclopedia of Sustainability Science and Technology.* New York: Springer.

Naiman, R.J. and Décamps, H. (1997). The ecology of interfaces: riparian zones. *Annual Review of Ecology and Systematics*, 28, 621–658.

Rose, D.B. (1996). *Nourishing Terrains: Australian Aboriginal Views of Landscape and Wilderness.* Canberra: Australian Heritage Commission.

Schneiderman, J.S. (2017). The Anthropocene controversy. In R. Grusin (ed.), *Anthropocene Feminism.* Minneapolis, MN: University of Minnesota Press, pp. 169–186.

Simmonds, H. and Gazley, A. (2018). Service ecotones: the complex boundary zones of service (eco) systems, *Journal of Service Theory and Practice*, 28(3), 384–404.

Springgay, S. and Truman, S. (2018). *Walking Methodologies in a More-Than-Human World: Walking Lab.* London: Routledge.

Taylor, A. and Guigni, M. (2012). Common worlds: reconceptualizing inclusion in early childhood communities. *Contemporary Issues in Early Childhood*, 13(2), 108–119.

Turner, N.J., Davidson-Hunt, I.J. and O'Flaherty, M. (2003). Living on the edge: ecological and cultural edges as sources of diversity for social–ecological resilience. *Human Ecology*, 31(3).

Weissteiner, C.J., Ickerott, M., Ott, H., Probeck M., Ramminger, G., Clerici, N., Dufourmont, H., and de Sousa, R. (2016). Europe's green arteries: a continental dataset of riparian zones. *Remote Sensing of Environment*, 8, 925. doi:10.3390/rs8110925.

Wojcieszek, J.M., Nicholls, J.A., Marshall, J. and Goldizen, A.W. (2006). Theft of bower decorations among male Satin Bowerbirds (*Ptilonorhynchus violaceus*): why are some decorations more popular than others?, *Emu: Australian Ornithology*, 106(3), 175–180, doi:10.1071/MU05047.

Zalasiewicz, J., Williams, M., Steffen, W. and Crutzen, P. (2010). The new world of the Anthropocene. *Environmental Science and Technology*, 44, 2228–2231. doi:10.1021/es90311.

2 Rivers of the Anthropocene

Introduction

This chapter flows from the intimate local of the Hawkesbury Nepean riverlands at Emu Green into the interconnected matrix of rivers and oceans of our blue planet to ask how are the major iconic rivers of the world becoming in the age of the Anthropocene? I do this digitally, lying on the red sofa in my house that overlooks the riverlands, flowing with the *Rivers of Life* television series. Watching episode after episode over and over again, I am enmeshed in their becomings as the rivers' images, sounds and stories wash over me until eventually I produce handwritten skeletal notes. These notes function as mobile images forming into words that are transformed into each iconic river's own river text. Further searches in the digital world of Google satisfy my desire to enter even more into becoming with the world's major rivers. Each river invites its own distinctive embodied knowing and poses questions that flow throughout the chapters of the book, from losses of the Nile, to the sacred materialities of the Ganges, the destruction and salvation of the Yangtze, the creations of the Mississippi, the transformations of the Rhine, and the multiple entwinements of the Amazon.

The Nile: river of conflict and loss

> ***Rivers and life***
> The Nile
> Longest river in world
> rises in heart of Africa
> White Nile from Lake Victoria in Uganda
> Blue Nile in mountains of Ethiopia
> lifeblood of greatest civilisation on earth.
> Egypt's rise due to historic right to own
> and control the Nile
> the eternal fertility of the soil
> a power unparalleled in human history
> replenished by annual deposits
> of rich and glorious mud

producing
vast abundance of food.
Ethiopia feeds the Nile
but is not able to feed herself.
Planning large dams to stop flow of water to Egypt.
Water wars will only be avoided
if every country sees the River as gift for all and prize for none.

Water wars

In 2002, in her book *Water Wars*, Vandana Shiva forecast with uncanny accuracy the events to come for the Nile River. As the world's longest river, the Nile is shared by 10 African countries including Ethiopia, the Sudan, Egypt, Uganda, Kenya, Tanzania, Burundi, Rwanda, the Democratic Republic of Congo, and Eritrea. It has long been a source of water conflict. The White Nile arising in Burundi, and the Blue Nile originating in Ethiopia have led to historical conflict between Egypt, Ethiopia and the Sudan. In 1958, Egypt began building the Aswan Dam, displacing 100,000 Sudanese residents, but the Sudanese were placated with the promise of more water. The water needs of all 10 countries are destined to multiply exponentially as the combined population grows from 245 million in 1990 to an estimated 859 million by 2025 (Shiva, 2002).

Hydropolitics

Egypt has historically dominated the Nile because of its bargaining and ideational power, economic development, military might, political stability, and access to external political and financial support. These enabled Egypt to influence and construct the Nile discourse (Mahlakeng, 2017). The recent emergence of China as a progressive financier to the global dams industry represents a shift for the international funding of hydraulic infrastructure as China becomes a major exporter of hydropower projects to the Nile region. Chinese companies and financiers have been involved in several dams in upstream countries challenging Egyptian dominance of the Nile. In the Great Lakes, Kenya has targeted the Nile to irrigate some 180,000 hectares of land. Uganda has followed suit, using the Nile to irrigate some 220,000 hectares and is also keen to exploit the Nile to generate power. The Karuma Dam is the country's biggest hydro-infrastructure project. These projects are all funded by China (ibid.).

The outcomes for the people of the river are dire. The Merowe reservoir, with a length of 174 kilometres displaced more than 50,000 people from the fertile Nile valley to an arid desert location. The Kajbar Dam displaced 10,000 people. Gibe III is the most destructive dam under construction in Africa. The project would condemn half a million of the region's most vulnerable people to hunger and conflict. In the entire Nile River Basin one million hectares of fertile land has been destroyed because of hydropower stations and dams (ibid.). There is great loss in these accounts of the damming of water, loss of land, livelihoods, places, and loss of stories.

Searching for what is lost

Simon Schama says we do not need yet another explanation of what we have lost, but an exploration of what we may yet find. For him, it is Egypt's fluvial myth of the cyclical death and rebirth of the waters that he seeks and finds in the legend of Osiris and Isis, lost when the island of Agilkia was flooded in the construction of the Aswan Dam (Schama, 1996, 14).

Mythical stories embody the representation of the deepest cultural truths about water in language. The 'great fluvial myths of the Nile' flow with cultural meanings about the transformation of the Nile basin in the cyclical return of the waters: 'in the late spring, hope, prosperity, and verdure returned to the basin of the Nile, born of the embrace between the moist Osiris and the earthy Isis' (ibid., 257). He writes about how mythical stories and their rituals can be re-created in the river's flow and its places:

> It was a modern embankment built to contain the rising waters until cultivation had prepared their fields to receive the floodwaters. A little truncated cone of earth, customarily known as 'the bride of the earth', a miniature Isis pyramid, had been built beside the dike and propitiatory offerings of millet and corn were strewn over it in offerings to the goddess of fertility. Some traditions called for a virgin, bedecked in muslin and flowers, to be thrown into the river to reenact with her body the union of the fertile earth goddess with Osiris.
>
> (ibid., 306)

Vandana Shiva too believes that 'Sacred waters carry us beyond the marketplace into a world charged with myths and stories, beliefs and devotion, culture and celebration', and that 'each of us has a role in shaping the creation story of the future' (Shiva, 2002, 139). To do this means creating new stories of water.

Touching and being touched by the Nile

The story of the Nile River fills my body with dread, a fear of the coming water wars, with water shortages predicted to be the greatest challenge of the Anthropocene. Of all the rivers it raises the most confusion, feelings awash with horror at the loss of water throughout the ages. It is only in the flow of its sacred stories that I can finally touch the Nile and allow the Nile to touch me. New stories are possible, stories that embody all that is within the river's life today. The Nile flows into Chapter 4 'Walking the songlines of the singing painting river', located in the loss of the Aboriginal stories songs and ceremonies that once echoed through the river's valley, and explores how new songlines can be formed.

The Ganges: river of sacred materiality

Rivers and life
The Ganges
Most revered river on earth
earthly form of goddess Ganga
the mother who grants eternal salvation.
Long ago the Ganges agreed
to fall to earth to redeem humankind
Lord Shiva caught Ganga in his hair at Gangotri
where Ganga is born in glaciers
thousands of pilgrims
make the long journey at time of ice melt
to bathe in cold flow of river and cleanse of sins.
People come to die at Varanasi
the most sacred place
and most polluted.
Ganga's spiritual and physical pulse
supports half a million people
pollution threatens river of life
to become river of death.
Over one million young children
poisoned each year by the River.
If power of love can save this river there is
hope yet as no river is as much loved as the Ganges.

A sacred river

All published writings about the Ganges address the Mother Ganga as a symbol of faith, hope, culture and sanity, as well as a source of livelihood for millions since time immemorial (Kumar, 2017). In this lies the paradox at the heart of the Ganges which I trace through these readings. On the one hand, Ganga is the most sacred river for Hindus, the special faith and respect for Ma (Mother) Ganga in India are as old as Indian culture itself. These are amply reflected in such ancient Indian scriptures as the Vedas, Puranas, Mahabharata, Ramayana and several others (ibid.). The very sight and touch of Ganga are so powerful that all the sins of many past births are destroyed. At the same time the Ganges River supports the practical needs of daily life, including its function as a waterway to carry away every kind of waste. These two things, the sacred, and the material, are intimately interconnected in the paradox of Ganga's stories.

Materiality of Ganga

The Ganges drains an area of over a million square kilometres. Its extensive basin accounts for one-fourth of India's water resources and is home to more than 407 million people, or about one-third of India's population. The Ganges basin, with its fertile soil, is a significant contributor to the agricultural economies of both

India and Bangladesh. By supporting agriculture, animal husbandry and fisheries, tourism, river-based trade and transport, the river contributes significantly to the livelihood, food and nutritional security of about one-third of Indian and two-thirds of the Bangladeshi population (ibid., 8).

In this lies its paradox, the fertility of the Gangetic plains is what enables the development of massive cities supporting large populations along its banks. The city of Varanasi, for example, the most sacred of all, has grown from 200,000 people to more than 3.6 million since the sewage system was built. All kinds of wastes including sewage inflow, industrial waste, animal carcasses, unclaimed human dead bodies, and other biodegradable and non-biodegradable wastes are dumped in the river (Kumar et al., 2012). Faecal coliform counts of more than 4 million/100 ml have been recorded at the downstream end of the city. Many studies have shown that there is a significant relationship between water-borne diseases and the use of the river water for bathing, laundry, washing of utensils, etc. (ibid., 343).

Materiality of sacred water

The writings about the Ganges Rivers straddle these two impossible realities: the sacred nature of its waters, and the extreme levels of pollution. Gangajal, or Ganga water, embodies this paradox within its very essence. Gangajal is one of the most important ingredients in Vedic rituals. Gangajal is held to be so sacred and holy that no Hindu ever dares to dishonour it and is obliged to speak the truth while holding the water. It is also the water of the Ganges that carries all forms of pollution from rubbish, sewage waste and toxic chemicals to human and animal carcasses (Kumar, 2017).

A remarkable recent finding about the presence of bacteriophage in the Ganges River is located in this paradox. An earlier study identified that the presence of bacteriophage, a virus that reproduces inside bacteria, was effective against the cholera virus in the early days of British colonisation. A current study has found the presence of bacteriophages which operate against putrefying and pathogenic bacteria in the waters of Ganges, even at Gomukh, its origin (Khairnar, 2016). The physical property of water at the Gomukh is unique, as loads of sediment gush out in force in the melting permafrost. The bacteriophages trapped at a much earlier time scale in the Himalayan permafrost are thought to be released gradually with the melting permafrost, thereby making a seed source of bacteriophage at Gomukh (ibid.).

Ganga's sacramental materialities

Watching the *Rivers and Life* episode of the Ganges, I become embodied in the flow of its waters from source to sea. I dwell in the sacred story of Ganga's fall from heaven, caught in Lord Shiva's hair at Gangotri, visit the source of the Ganges at Gomukh, the cow's mouth, with its frozen water, releasing cleansing bacteriophages. I go to the place where people come to die at Varanasi, and I hold galangal, the sacred water in my hands. As I have these experiences of the

Ganges River, I wonder how it is that a river becomes sacred, a wondering that continues in Chapter 6 where I explore bedrock's sacramental becomings.

The Yangtze: river of destruction and salvation

Rivers and life
The Yangtze
Key to China's economic growth
power to make or break
fertilises rice fields with rich sediment
gives and takes away
5 million homes lost to flooding.
3 Gorges Dam
quest to control nature in pursuit
of progress
most ambitious and controversial engineering
project in history of planet
visible from space built on faultline
in earthquake zone
tames Yangtze from wild to navigable
enables China's industrial ambition
200,000 cargo ships into heartland
revitalising inland provinces with
enough power to support their growth
into largest mega cities in world
50 million lives submerged
25 million tonnes waste discharged into river
every year
life-giving silt blocked
fish stock depleted, plankton reduced
Yangtze trapped
in poisonous cycle of growth.

Destruction and salvation

Much has been written about the controversial Three Gorges Dam, especially as time passes since its construction in 2003. It is often difficult to determine how valid the contributions are from the West in judging China's actions from the outside. Chinese authors, Yang and Lu (2013) review the benefits and impacts of the controversial mega-dam over the ten years since its construction with regard to future environmental protection, offering a useful way of moving beyond this western-centric focus.

Destruction I

The damming of water has resulted in lower water levels in the downstream channel and less water storage in adjoining riparian lakes. Stagnant water in the tributaries has elevated pollution levels. The riparian ecosystem has been significantly disrupted with the destruction of spawning grounds and the obstruction of migration routes, resulting in the loss of a large number of fish species and freshwater cetaceans. The latest investigation in 2012 shows that the population of finless porpoises in the Yangtze River basin has declined to approximately 1,000 (ibid.).

Salvation I

Yang and Lu write about the very substantial benefits of the Three Gorges Dam (ibid.). These include the salvation of 15 million people and 1.5 million acres of farmland from regular flood disasters. The cumulative power generation has surpassed 629 billion kilowatts and saves nearly 600 million tons of carbon dioxide emissions into the atmosphere, a very significant benefit considering the impact of climate change on the glacial sources of many of the world's rivers. Less convincing, but still an important factor, they include the improvement of waterway navigability, with the Yangtze continuing to be the world's busiest cargo-bearing river, with 1.5 billion tons of throughput in 2010, important to China's growing economy (ibid.).

Destruction II

As China soared to new economic heights, the *baiji* dolphin was ill-equipped to fend for herself. Fishing nets and boat propellers came after her, increasingly toxic waters poisoned her food, human fishers took a hefty share of what was left, and the Three Gorges Dam sealed her fate. By the beginning of this century she succumbed, becoming the first dolphin driven to extinction by humans (Jin, 2017).

Salvation II

The story of Princess Baiji, the finless dolphin, flows deep under the Yangtze River's busy water highway (ibid.). Princess Baiji was to be married to a man she didn't love. When the princess refused, her father pushed her into the Yangtze River. Princess Baiji drowned, but the river took pity on her and reincarnated her into a dolphin. For millennia, the so-called Goddess of the Yangtze pulsed her slender body through the river's currents.

Destruction III

In yet another twist in this winding river tale, the story continues that the myth of invincibility of the finless porpoise is just that. In a river once flush with thousands of porpoises and dolphins, today only around 1,000 finless porpoises still navigate the Yangtze's twists and turns. The finless porpoise now is critically endangered with less than 10 years to stave off extinction.

Salvation III

The story continues in a tribute to the life force of the river. Remarkably, the finless porpoise, Baiji's less-storied cousin, still swims the river. The Yangtze finless porpoise is fleshy and rotund, its mouth fixed in a permanent grin. Nicknamed 'river pig' by locals, it has long been rumoured to repel any weapons hurled at it. Thus far, this eternally smiling kung fu master of the river has survived the environmental assaults that conquered the *baiji* dolphin (ibid.).

Salvation IV

The ongoing story of Princess Baiji, the finless dolphin and her cousin the finless porpoise, continues in the context of China's recognition of environmental protection as an urgent issue. Environmental protection, rather than big development projects, has become a dominant focus of river development plans. China's most comprehensive water policy to date aims to significantly improve the water in seven key rivers, including the Yangtze, to 70 per cent 'excellent or good' by 2020 (ibid.). The World Wildlife Fund (WWW) enters the story and engages corporations that have the largest environmental impact. These include textiles, soft drink and alcohol manufacturers, and others who depend on the continuing environmental integrity of Yangtze ecosystems for their businesses to remain viable. Not far from Dongting Lake, a programme jointly sponsored by the Chinese government, the WWF, the Coca-Cola Company, HSBC, and others is offering the finless porpoise a lifeline. Porpoises are being bred and protected in small U-shaped lakes called 'oxbows' in the Tian-e-Zhou Oxbow Reserve. The hope is that the porpoises will be able to return to a cleaner Yangtze River.

Destruction and salvation

Witnessing the destructive power of the massive traffic on the Yangtze River, after the construction of the Three Gorges Dam, is crushing. The sense of the endless movement and churning waters as cruel vessels thrash their way through everything in their pathway becomes intolerable. It was not until I read the mythical story of Princess Baiji, the rare finless dolphin, and her cousin, that I could alternatively float and submerge with this paradox of destruction and salvation. The rise and fall of the finless dolphin and her cousin, the finless porpoise, came to represent the entanglements of human and the world's rivers in this time of the Anthropocene. The Yangtze's question as it flows into Chapter 7, 'The river's crossings' is, how can humans become other to themselves to enable them to dwell in the life of the other of human becoming animal and animal becoming human?

The Mississippi: river of creation

Rivers and life
The Mississippi
made America
the richest and wealthiest nation on earth
and most impoverished.
Land of contrasts
north/south rich/poor white/black
Delta richest
cotton-growing area in America
African American slave trade
blues comes from holler songs
calling and answering in rhyme and song
make day easier to pass by
low down dirty shame blues and
up tempo blues [1]
Mississippi
Rocked a nation
Blues, rock and roll, jazz, gospel
flowed into America
on muddy waters of Mississippi
and out again into the world
stars spangles and suffering made America
what it is today.

Creating music

> Much of what is profoundly American, what people love about America has come from the Delta.
>
> (National Park Service, 2017)

The Mississippi is the longest river in the United States, flowing 3,730 kilometres from its source in Lake Itasca in northern Minnesota to its delta in the Gulf of Mexico. From its source to the sea, the 'muddy' river carries rich deposits of silt which produces one of the richest landscapes in the country. Its broad, alluvial valley reaches from southern Illinois to the south-eastern tip of Louisiana, covers more than 90,000 miles of rivers and streams, more than 3 million acres of land. The Delta's cultural traditions are as rich and diverse as its natural resources. It is a land of converging cultures with a unique complexity and density of history, prehistory, and cultural expression. Over the centuries, American Indians, French, Arab, Spanish, African, German, English, Irish, Scots-Irish, Jewish, Italian, Chinese, Mexican, and Southeast Asian peoples have established and maintained their distinctive ethnic identities (ibid.).

While the Delta's musical styles are diverse, it is the blues music of the lower Mississippi River Delta that is best known around the world. Developed by people engaged in struggle, the blues are rooted in African music that evolved from field hollers, the work songs of slaves that carried deeply layered, coded messages. Infused with the spirit and struggle that produce creativity and resilience, the blues tell stories of frustrated love, broken homes, and other miscues of an oppressed and displaced people. They came into being in the cotton fields by musicians with no formal training, but an ear for the rhythms of their daily lives (ibid.).

Creating art

The Mississippi River's many other forms of creation are documented by a number of writers in an edition of *Open Rivers* (Nunnally, 2015) that focuses on *Rethinking the Mississippi*.

Wes Modes' project, 'Secret History', is the culmination of the artist's dream. He built a replica of an early twentieth-century shanty boat from scratch and floated down the Mississippi River, listening to the stories of its river people. In stepping into the river of history, he heard from scientists, homeless people living under bridges, merchants, bar owners, river rats, artists, researchers, fishermen, boaters, and people who have lived within sight of the river every day of their lives. He came to understand that we are all river people. As humans, we depend on, and indeed are made of, water. Like the river itself, river knowledge is both deep and wide (Modes, cited in ibid.).

Artist Bernard Williams was compelled by the river's ability to claim lives. This became the subtext for a sculpture project concerned with invisible root systems and the river's meandering history. Like Wes Modes, he engaged with the waters in his own makeshift raft, but he created a sculpture, 'Ghost Trails', from the experience. Ghost Trails echoes old river routes no longer taken. It recalls river baptisms, suggesting the potential and need for spiritual and mental conversion. As a river artist, he is the convert stirred to new awareness by the complex beauty and enlightening revelation of the river's meandering and liquid thought (Williams, cited in ibid.).

Creating stories

For Angela Miller, the most profound chroniclers of the Mississippi River were novelists and writers. She describes the river of Herman Melville's *The Confidence Man* as a vast ungovernable and godlike force of nature that provides the backdrop for the flickering shadow play of human ambition and deception. In Mark Twain's *Huckleberry Finn* she finds a utopian space of possibility in the river where one might escape social evils, petty feuds, and corrupt institutions. *Huckleberry Finn* takes her into the heart of darkness as Twain's Mississippi connects the industrial North of free labour to the slave plantations of the South, binding the two together through an infernal pact. Meaning emerges in these stories from the intersection of the material and the conceptual through which both river and literature become more real (Miller, cited in ibid.).

John Afinson finds the Mississippi gathers stories like tributaries, growing culturally stronger and deeper with each one. These stories include contributions to American literature, art, music, and to the national narrative, the national story line. For him, they are also cautionary tales, to know the river is to care for it, but every year cultural resources are lost that are the tangible reminders of those who lived along and used this river in the past. Habitat of migratory birds, aquatic and terrestrial plants and animals is being increasingly lost and knowing the river's stories means working to slow and reverse these trends (Afinson, cited in ibid.).

Creating with the Mississippi

Immersed in its creations, I come to know the Mississippi in a practice of art, story, music and song through which I can become part of the river's creations in some small way. Every day, while coming to know the Mississippi I listen to, and sing along with, gospel and blues, feeling with the bittersweet rhythm of the holler songs resonating through my body. The Mississippi asks, how can we create with the rivers of the world as it flows into Chapter 9, 'Global materialities'? I make art creations with litter from my river and exchange stories and artmaking with waste from the Mississippi with my collaborator in Memphis, Tennessee, as we form new identities as Feral Susan and Emu Girrl, exchanging letters that meet on the waste/d Henderson Island.

The Rhine: river of transformation

Rivers and life
The Rhine
The great artery of Europe
1,300 kilometres from source to sea
glacial source Swiss Alps
Switzerland, France and Germany
ending at sea in Rotterdam.
Iconic river of culture and tradition
patriotism, nationalism, monuments
Lorelei Rhine maidens, folklore.
Crucial to Europe's economic success
sea port delta
one of busiest ports in the world
Tamed waters
straightening the Rhine,
canal cutting through Europe
sharing goods and cultures
tyranny war marching armies, thunder of bombs
persecution, concentration camps, murder, death
Great success story reducing pollution
Eels and salmon once again plentiful
in the great artery of Europe.

Transformation, from pristine to industrial

The challenges for the Rhine are similar to those in all other major rivers in its encompassing of many national constituencies, including Switzerland, France, Germany, the Netherlands, Belgium, Luxembourg, Liechtenstein, Austria and Italy. For centuries the transport of bulk goods over the Rhine was of paramount importance to the economies of Western Europe. And yet, the Rhine differs from all others. International cooperation on the protection of the Rhine drastically changed in November 1986, after the Sandoz accident. During the dousing of a fire in a warehouse of Sandoz AG in Basel, Switzerland, about 15,000 cubic metres of water, mixed with highly dangerous chemicals, flowed into the Rhine, forming a 70 km red trail that moved slowly downstream from Switzerland, through France, Germany, and the Netherlands. Hundreds of thousands of dead fish and waterfowl washed up along the banks of the river. All drinking water processing companies using Rhine water shut down. From then on, international policies led, rather than lagged behind, domestic efforts (Verweij, 2017).

Transformation policy processes

Some environmental problems remained or re-emerged and, since 2000, several new initiatives have been launched. The most powerful of these is the EU's Water Framework Directive (WFD) of December 2000. The initial objective of the directive was to achieve, by 2015, good ecological quality for all EU waters with the help of river basin management. Under the directive, all 26 EU member states are obliged to identify their river basins and set up river basin districts. For each district, a six-yearly River Basin Management Plan had to be written. The directive also specifies which biological, physical and chemical parameters should be measured, how deviations from reference conditions should be enumerated, how monitoring programmes should be designed, in what manner results should be presented, and compiled timetables for the completion of different tasks (ibid.).

Uncommon transformation

Actions resulting from these international policy directives resulted in the Rhine watershed being transformed from the 'open sewer of Europe' to being 'the cleanest river in Europe'. In the Rhine basin, upstream users helped downstream users, companies went beyond legal standards. The release of chemical pollutants and heavy metals reduced, the dumping of salt into the river discontinued, flood protection systems improved, warning systems against accidental spills were put in place, run-off of nitrogen and other harmful substances from agriculture decreased, monitoring of the Rhine's water quality improved, passageways for migratory fish were built and ecosystems were at least partly restored. As a result, oxygen levels have rebounded and many species of fish and other animals have begun to return to the river (ibid.).

Mythical transformations: the Lorelei

The mythical legends of the Lorelei represent the living links between the intangible of human relations with rivers, the rivers themselves, and the stories through which humans come to express this complexity. The legends similarly change over time with the pulse of the river.

Transformation from Water Spirit temptress to Forsaken Bride

In earlier traditions the Lorelei is represented as an undine (water spirit), combing her hair on the Lorelei-berg and singing bewitching strains to lure mariners to their death. An old soldier Diether and his little band of warriors set out to capture her. Instantly, a wild storm arose, and the river, overflowing its banks, foamed right up to the summit of the Lorelei Rock. Three white-crested waves mounted the steep rock, and into the hollowed trough behind them. The Lorelei stepped into the waves, to be whirled out into the river in her chariot. In mid-stream the undine sank beneath the waves, the spell was broken, the waters subsided, and the captain and his men were freed to return home (Spence, 1995).

In a later legend the Lorelei was a maiden of such unsurpassed beauty that every gallant on whom her eye rested fell hopelessly in love with her. The truth was, she had given her heart into the keeping of a young knight who, after plighting his troth with her, had ridden away to the wars. Years had gone by and Lorelei thought he had perished or taken another bride and forgotten her. She was besieged by an ever-increasing band of suitors and it began to be rumoured that she was a sorceress, generally condemned to death by burning. Because of her great beauty, she was sentenced to a convent, but on the way they passed a light barque sailing up the river, and as she gazed on it Lorelei uttered a loud cry, for there in the bow stood her truant lover. The frail craft ran onto the rocks, Lorelei saw her lover's peril and, calling his name, leapt into the tide. Nothing more was seen of the lovers; together they sleep the sleep of death beneath the waters of the Rhine (ibid.). The legend of the Lorelei keeps alive the possibility of cyclical transformation in human river entanglements.

Pulsing with the great artery of Europe

The lifefulness of the regeneration of the Rhine pulses through my body with a renewed sense of energy and hope. I feel the cleansing water wash over me as fish and other aquatic creatures return to swim in its cleansed waters. The Rhine's story of its cyclical transformations reminds the world that the rivers are always greater than their human population, keeping alive the possibility of transformation even in the worst of conditions. The Rhine flows into Chapter 8, 'Regeneration: of trees, weeds and tender intimacies', asking the question, how can our rivers be transformed, as we sing the song of the changing seasons of the bushcare regeneration of the riparian zone at Emu Green?

The Amazon: river of entwinement

Rivers and life
The Amazon
a force of nature
glacial source high in Andes in Peru
6,500 kilometres long
more water than any other river
on the planet
wild powerful never tamed
700 million people living on banks
can't control its force must adapt
thousands plants and animals uncatalogued
quarter of all pharmaceutical drugs come from
delicate ecosystem of the Amazon.
last tropical rainforest on Earth
trees are water towers of the Amazon
draw water from ground release
into atmosphere falls as rain
Aggressive property development destroying
jungle to point of no return.
6 football pitches cut down every minute.
destroying indigenous communities and livelihoods
soybean, logging, cattle deforestation drought forest fires
for how long the last great untamed river on earth
will continue to run wild remains to be seen.

Entwinement of Ribereños

The first and most immediate intertwinement is to be found in the lives of the Ribereños, the river people, who inhabit the riverbanks of Amazonia. The Ribereños live on the rich floodplains that annually undergo drastic environmental change. The river is their life and is used for washing, bathing and fishing, and the water for cooking. Since there are no roads in the forest, the rivers are the roads and highways. Everyone travels by dugout canoes. Farming is the major focus, but fishing, hunting, extraction of forest products are also sources of sustenance. The people have great knowledge of forest plants and the forests themselves reflect the ethnicity of the Ribereños because of the way they select and utilize wild species (Talk to an Expedition Specialist, n.d.).

Socio-ecological entwinement

The notion of 'Amazonia' appears, not only as the river but also as the river and its ecosystem as a whole, which has 'long been the focus of worldwide attention for its remarkable biological and cultural diversity and for its role in regulating global biogeochemical and atmospheric cycles' (Tucker Lima et al., 2016, 2). At the same time, the region has been increasingly viewed within recent decades as a frontier ripe with opportunities for economic growth and large-scale development. These two imperatives are closely intertwined, as national and international policies seek to advance Amazonian development, and environmental interests struggle to conserve the region's unique rainforest ecosystem. Meanwhile, new growth policies increasingly focus on how to reconcile conservation and development (ibid.).

Entwinement of global markets, soy production and deforestation

Other writers offer an account of the relationship between market forces and land clearing over time. Land-use transitions differed dramatically between the periods 2001–2005 and 2006–2010. The first period corresponded to a boom in cropland expansion (primarily soy), with the area planted in soy doubling from 3 to 6 million hectares and production increasing by 85 per cent or 8 million tons. A third of that increase in area (~1 million ha) and production (~3 million tons) occurred in the Amazon forest biome, where the planted area more than tripled during the same period. Rising demand for soy was primarily related to export markets for animal fodder in Europe and Asia (Macedo et al., 2012).

From 2006 to 2010, deforestation in the Amazon declined dramatically, particularly in Mato Grosso, Brazil's leader in soy and beef production, responsible for 31 per cent of the nation's soy production and over 13 per cent of its cattle herd in 2009. From 2000 to 2005, it also led in deforestation, accounting for 40 per cent of deforestation in the Brazilian Amazon. In the ensuing years, deforestation in Mato Grosso declined substantially, reaching an estimated 850 km by 2010, just 11 per cent of its historical average. These declines in deforestation coincided with fluctuations in commodity markets and the implementation of several high-profile policy initiatives aimed at restricting credit for deforesters, improving monitoring and enforcement, and excluding deforesters from the supply chains of major exporters (ibid.).

Climate entwinement

Amazon ecosystems encompass 10–15 per cent of land biodiversity. The abundant rainfall of about 2.2 metres per year makes the region an important heat source for the atmosphere. The Amazon generates an estimated 210,000–220,000 metres of river discharge, which is about 15 per cent of the freshwater input into the oceans. It stores an estimated 150–200 billion tons of carbon; and it presents a mosaic of ethnological and linguistic diversity (Nobre et al., 2016).

A number of complex problems threaten the geopolitical, environmental, social, and economic stability: the links between global food and energy markets; the unsustainable depletion of natural resources and biodiversity stocks; the increasing water insecurity around the world; and, above all, the urgent need both to decarbonize the energy systems of the world to avoid catastrophic climate change and to adapt to unavoidable climate change underway. The scale and reach of the risks associated with climate change, together with their potentially irreversible nature, make this probably the greatest market failure and the starkest example of a 'tragedy of the commons' the world has ever seen. Reducing deforestation could minimize these impacts as well as mitigate emissions of greenhouse gases (ibid.).

Dwelling in intertwined intimacy and immensity

I float in the space between the Amazon River, the largest river in the world, and its rainforest trees, inseparably intertwined with the fate of the planet. The rate of clearing and burning of rainforest continues when it is only trees that can save the planet from the catastrophic consequences of anthropogenic global warming. The Amazon inserts itself into Chapter 10, which explores 'Life and death in the Anthropocene', and the very phenomenon of anthropogenic climate change that threatens the life of the iconic rivers of the world at their sources in the great glaciers of the planet.

Anthropocene entanglement and glacial sources

> The Anthropocene represents a new phase in the history of both humankind and of the Earth when natural forces and human forces became intertwined so that the fate of one determines the fate of the other. Geologically this is a remarkable phase in the history of the planet.
>
> (Zalasiewicz et al., 2010, 2231)

The human is already always entangled with the iconic rivers of the world: in the sacred of the Ganges; the creations of the Mississippi; the regeneration of the Rhine; the conflict of the Nile; the salvation and loss of the Yangtze; and in the entwinements of the Amazon.

The rivers provide food through spreading rich silt, human populations boom producing human waste and pollution; legends and mythologies emerge, are submerged and reappear. Policies are formed and reformed in the ever-changing cycles of human-river relations. All rivers provide microclimates for their human inhabitants, and the powerful Amazon through its entwinement with rainforest has a global planetary influence on climate. We talk of the Anthropocene as the time of the sixth great extinction, but what of the extinction of the rivers themselves as the glaciers at their source disappear?

Glacial sources, climate change, and the Anthropocene

Deleuze noted that elements enfold to produce a human 'figure' that comes to express the world. A human subject is but an instance of subjectification that becomes an expression of a chaotic world that far exceeds the confines of any humanist ontology, and emerges as a fold of the elements in their various forms: water, air, dust, waves, particles, fog, glaciers, sunshine, etc. As a fold, the human subject emerges as a momentary pleat in a material 'chaosmos', which Deleuze defines as a 'dark backdrop' (Deleuze and Guattari, 1987, 87). The classical elements, air, water, earth, fire are not distinct from each other, but are pleated in new ways through recursive and erratic geological and biological cycles and processes (Piotrowski, 2019).

All of the iconic rivers are connected to the new epoch of the Anthropocene through their source in retreating glaciers, but the relationship between the rivers, glaciers, climate change, and climate is as variable as the rivers themselves. For all except the Mississippi, the impact of glacial melt is noted in the literature, but even the Mississippi derives its typical glacial formation from the Great Ice Age.

The Upper Mississippi River Valley was shaped during the glacial stage of the Great Ice Age when the melting of enormous ice sheets, up to 5,000–10,000 feet thick and covering hundreds of thousands of square miles, released tremendous amounts of water, forming huge glacial lakes. The glaciers then left their mark by carving the spectacular Mississippi River that cuts through the landscape with the torrents of water draining from the melting ice mass. The largest of the glacial lakes, Lake Agassiz, is the glacial source of the Mississippi River today (Middleton, Nissen, Wisconsin DNR et al., n.d.).

The Gangotri glacier, the place where the Goddess Ganga touched down on Earth, is the site of the Ganges' most sacred glacial relationship. Nearly 4,000 years ago, the glacier was about 18 km longer with a terminus at a waterfall over carved granite. The rock carved by the flow of the river and ice had the appearance of water emerging from a cow's mouth, thus giving it the name Gaumukh in Sanskrit. Through time, the glacier receded up the valley, but the name Gaumukh remains as the place from which the sacred Ganga emerges from the Himalayan Mountains, regardless of its exact position as the glacier has changed (Braasch, n.d.).

The Amazon relationship with its glaciers is as intertwined as with its rainforest. The mass balance of the Andean ice masses that flow towards the Amazon drainage basin is strongly dependent on the Amazon hydrological cycle. The main source of their snow precipitation is air masses bringing water from the Atlantic Ocean, water recycled through the rainforest several times. Changes in this air flow due to climatic warming are the main drivers of the rapid retraction and the disappearance of high-altitude ice fields and glaciers in the tropics. Unlike glaciers in temperate regions, where precipitation on the glacier occurs during the austral winter at low temperatures, in the tropics it happens during the warmer austral summer months (Ribeiro et al., 2017).

Seventy per cent of the glaciers in the Yangtze River basin are found in the region above the Tongtian River; the total ice volume of the glaciers in this region is 100,414 cubic metres. The statistical GPS study found that, in general, the glaciers in the source region of the Yangtze River decreased 126,33 square kilometres between 1977 and 2009, a total of 11.84 per cent pf the area between 1977 and 2009. The important quality of Yangtze glacial relationships, however, is their broad socio-ecological meaning. Glacier meltwater is both an important water source for rivers and an important resource for nomadic herders. It is also the main source of domestic water in the source region of the Yangtze River and the freshwater resource relied on by the wild animals of the plateau.

The Rhine's glacial relationship tells of its seasonal fluctuations. More than a quarter of the water that flows from the Rhine into the Mediterranean in August has its origin in alpine glaciers. At its mouth in the Netherlands, 7 per cent of the water in the Rhine is glacier run-off. The alpine glaciers are shrinking and will continue to do so but the problem is not that there would be a lot less water overall, but that it would be there at different times. Whereas the water coming from the sky as snow in winter is now stored in the glaciers and released in the summer, in the future it will enter the river systems much more quickly. This could mean floods in spring as the snow melts, and low water levels and even drought in the summer.

Finally, the extraordinary glaciers of the Nile are located in the Rwenzoris Mountains that rise 5,000 m from the heart of Africa. The 'Mountains of the Moon' in the heart of Africa remain virtually unknown outside of Africa. When they were first climbed in 1906, their mere existence caused a sensation in Europe. Luigi di Savoia, the Duke of Abruzzi, reached the peaks and returned home with dramatic photos, revealing a surreal landscape of snow and ice, looming over the rainforest below. Glaciologists estimated that the famous glaciers of the Rwenzoris had dwindled by more than 80 per cent in just a century, and predicted that the remaining ice would be gone in less than 20 years. The future of the mountains, their remarkable endemic biodiversity, and the Bakonjo people who call them home remains uncertain (Losin and Dappen, 2014).

It seems to offer a fundamental onto-epistemological shift to imagine the extinction of so many rivers in this time of human entanglement in the fate of the planet:

> If the elements are an integral force in the production of the notion of the species' emergence in time requires new forms of narrative [Darwin], and imaginative and ethical articulation, then the intensifying sense of the species' end makes a similar claim for rethinking 'our' processes of self-presentation and self-preservation. Rather than one more event within time, not only does the Anthropocene require us to shift our scale of narration away from human generations and history to species emergence and deep time; it also raises the problem of intersecting scales, combining the human time of historical periods (late capitalism, industrialism, nuclear power) with a geological time of the planet. This in turn requires us to open the classically feminist question of the scale of the personal.
>
> (Colebrook, 2017)

The world's rivers flow into and out of the oceans in the interconnected matrix of rivers and oceans that shape our blue planet, but it is only on the scale of the intimate local and personal that we can become with the immensity of the world's rivers.

The Hawkesbury (Deerubbin) and Nepean (Yandai) River is a system of interconnected waterways that encircles the city of Sydney, travelling 178 kilometres from its source in remote mountainous country south of Sydney until it meets the south Pacific Ocean at Barrenjoey headland in Sydney's north. It is not one of the major iconic rivers of the world, but it has its formation in the ancient Gondwanaland, and is the home of the oldest living culture in the world, the Darug, Dharawal Darkinjung, Eora, Gundungarra, and Kuringgai Aboriginal peoples. The river provides water for Sydney's 5 million people from the Warragamba Dam, built in the 1950s. Prior to the building of the dam, a series of major floods devastated the lower floodplains, carrying bridges, roads and houses away, and leading to the evacuation of thousands of people.

Emu Green on the Nepean River, about halfway along the river's course to where the Hawkesbury River begins at Yarramundi, is my river home. It continues to gather up all the elements, weather, soil, air, water, and all the living things that make up the complex ecologies of waterways, including its fish and aquatic creatures, insects, birds, trees, plants, and humans as an integral part of its whole. Together they produce new onto-epistemologies of the human through Australian Aboriginal eco-philosophies, and western posthuman and new materialist theories. It remains connected to all of the rivers of the world through the world's oceans, and the iconic rivers wind themselves in and out of the chapters, always on the move, changing with seasons, time of day, weather, and so on. The Hawkesbury Nepean river poses the final question in the becomings of the rivers of the world, it is the question of methodology, of how I can bring the infinitesimal moment by moment becomings of the riverlands to you, the question that informs Chapter 5: the riverlands' watery ways.

As all the rivers gather together at the end of this chapter, there is some final questions to ask about the element that belong to them all, it is about water. The rivers' last questions are, what are water's stories?, what does water do?, and how does water shape the human in its eternal flows through the bodies of the world? These are the questions that shape Chapter 3, 'Watery beginnings'.

Note

1 Honeyboy blues singer.

References

Amos, J. (2016). Geologists search for Anthropocene 'golden spike'. *BBC News: Science and Environment*. Available at: www.bbc.com/news/science-environment-37200489 (accessed 20 February 2017).

Barbezani, B., Labandera, E., Timpe, K. and Kaplan, D. (2016). A social-ecological database to advance research on infrastructure development impacts in the Brazilian Amazon. *Scientific Data*, 3. doi:10.1038/sdata.2016.71.

Barthem, M., Marques, P., Charvet-Almeida, P. and Montag, L.F.A. (2005). Soy moratorium impacts on soybean deforestation dynamics in Mato Grosso Brazil. *WIT Transactions on Ecology and the Environment*, 81, 615–625. Available at: www.ncbi.nlm.nih.gov/pmc/articles/PMC5408992/ (accessed 20 December 2018).

Braasch, G. (n.d.). World view of global warming: the photographic documentation of climate change. Source of the Ganges. Available at: www.worldviewofglobalwarming.org/ganges/ (accessed 20 December 2018).

Colebrook, C. (2017). We have always been Post-Anthropocene. In R. Grusin (ed.), *Anthropocene Feminism*. Minneapolis, MN: University of Minnesota Press.

Deleuze, G. and Guattari, F. (1987). *A Thousand Plateaus: Capitalism and Schizophrenia*, trans. B. Massumi. Minneapolis, MN: University of Minnesota Press.

Jin, J. (2017). Birth along China's Yangtze River: Follow a journey up and down China's Yangtze River and learn about a multi-layered effort to protect everything from rare porpoises to sustainable economic growth. *World Wildlife Magazine*, Fall.

Khairnar, K. (2016). Ganges: special at its origin. *Journal of Biological Research*, 23(16). doi:10.1186/s40709–016–0055–6.

Kumar, D. (2017). River Ganges: historical, cultural and socioeconomic attributes. *Aquatic Ecosystem Health & Management*, 20(1–2), 8–20, doi:10.1080/14634988.2017.1304129.

Kumar, S., Jha, P., Baier, K., Jha, R. and Azzam, R. (2012). Pollution of Ganga River due to urbanization of Varanasi: adverse conditions faced by the slum population. *Environment and Urbanization, Asia*, 3(2) 343–352.

Losin, N. and Dappen, N. (2014). Snows of the Nile: the search for vanishing equatorial glaciers. Available at: www.youtube.com/watch?v=szZ15fFdilw

Macedo, M.N., DeFries, R.F., Morton, D.C., Stickler, C.M., Galford, G.L. and Shimabukuro, Y.E. (2012). Decoupling of deforestation and soy production in the southern Amazon during the late 2000s. *PNAS*, 109(4), 1341–1346.

Mahlakeng (2017). China and the Nile River basin: the changing hydropolitical status quo. *Insight on Africa*, 10(1), 73–97.

Middleton, P., Nissen, R., Wisconsin DNR, et al.(n.d.).Glacial lakes and rivers form the Mississippi River Valley.Availableat:www.greatriver.com/Ice_Age/glacier.htm

National Park Service (2017). History and culture of the Mississippi: Lower Mississippi Delta Region. Washington, DC: U.S. Department of Interior. Available at: www.nps.gov/locations/lowermsdeltaregion/history-and-culture-of-the-mississippi-delta-region.htm (accessed 20 December 2018).

Nobre, C.A., Sampaio, G., Borma, L.S., Castilla-Rubio, J.C., Silva, J.S. and Cardoso, M (2016). Land-use and climate change risks in the Amazon and the need of a novel sustainable development paradigm. *Proceedings of the National Academy of Sciences of the United States of America*. https://doi.org/10.1073/pnas.1605516113 (accessed 20 December 2018).

Nunnally, P. (ed.) (2015). *Open Rivers: Rethinking the Mississippi*, Issue 1, Available at www.lsu.edu (accessed 20 December 2018).

Piotrowski, M. (2019). 'An atmosphere, an air, a life:' Deleuze, elemental media, and more-than-human environmental subjectification and education. *Environmental Education Research*, https://doi.org/10.1080/13504622.2018.1485134.

Ribeiro, R. R., Simões, J.C. and Ramirez, E. (2017). The Amazon glaciers. In D. Odone (ed.), *Glacier Evolution in a Changing World*. Intech Open. http://dx.doi.org/10.5772/intechopen.70490.

Schama, S. (1996). *Landscape and Memory*. New York: Random House.

Schneiderman J. (2017). The Anthropocene controversy. In R. Grusin (ed.), *Anthropocene Feminism*. Minneapolis, MN: University of Minnesota Press.

Shiva, V. (2002). *Water Wars: Privatisation, Pollution, and Profit*. London: Pluto Press.

Spence, L. (1995). The Rhine in folklore and literature. In L. Spence, *Hero Tales and Legends of the Rhine*. New York: Dover Publications.

Talk to an Expedition Specialist (n.d.). Available at: https://au.expeditions.com/destinations/south-america/amazon/the-experience/read-up-gear-up/staff-article9/ (accessed 20 December 2018).

Tucker Lima, J.M., Valle, D.Mateus Moretto, E., Mantovani S., Pulice, P., Lucia Zuca, N., et al. (2016). A social-ecological database to advance research on infrastructure development impacts in the Brazilian Amazon. *Scientific Data*, 3. doi:10.1038/sdata.2016.71.

Verweij, M. (2017). The remarkable restoration of the Rhine: plural rationalities in regional water politics. *Water International*, 42(2), 207–221, doi:10.1080/02508060.2017.1278576.

Yang, X. and Lu, X.X. (2013). Ten years of the Three Gorges Dam: a call for policy overhaul. *Environmental Research Letters*, 8, 041006.

Zalasiewicz, J., Williams, M., Steffen, W. and Crutzen, P. (2010). The new world of the Anthropocene. *Environmental Science and Technology*, 44, 2228–2231.

3 Watery beginnings

Introduction

This chapter, *Watery beginnings*, poses the questions asked in consort by all of the rivers of the world, what are water's stories?, what does water do?, and how does water shape the human in its eternal flows through the bodies of the world? In response to the rivers' questions, I travel with the flow of the Hawkesbury Nepean River, from the fresh water of my daily swim hole at Emu Green until it reaches the sea at Broken Bay where fresh water and salt water meet. Here in this edge place, new stories of water emerge: from water's beginnings of life on Earth; the beginnings of human life in amniotic fluid; the waters of first contact; the waters that carried my ancestors from Scotland; and the watery beginning of the mega city of Sydney in Warragamba Dam. Waters' stories traverse two very different time zones, held in the space between the geological beginnings of life, and the formation of the mega city as it exists today. Cradled in between are the memories of the watery beginnings of human lives on Scotland Island, the place where four of my babies were born. The time to write on Scotland Island is about recreating water's memory traces in body and mind, reliving the watery beginnings of life on Earth and the watery beginnings of human lives.

> ***Swim hole, 30 December 2018, 7.50 a.m.***
> Walking along narrow track worn
> by moving feet
> push aside curtain of draped balloon vine
> and clamber down rugged bank to reach
> three tall casuarina trunks
> temporary stopping place to launch
> into waist-deep water and cruise
> to ripple of current over rock where
> body launches into the flow of
> watery memory traces travelling all the way
> to the place where we lived so long ago
> at the meeting of fresh and salt water's beginning of
> life on Earth.

Water's beginnings of life on Earth

I swim my way through the *The Moth and the Flame*, the third episode in Brian Cox's BBC series, *The Forces of Nature*. Set on our blue planet, the planet of water, it is about the moment after 500 million years of no life, when the inanimate became the animate, the beginning of life on Earth. I am swimming, not in the world's rivers this time, but deep under the ocean in underwater caves, the most dangerous of expeditions, searching for the origins of life on Earth. Life, for Brian Cox, is 'just a temporary haven for the elements that make up the universe, a very special theatre from which biology emerged, the vital ingredients of life, hydrogen plus oxygen equals water, H_2O'. These vital ingredients flow through water because it dissolves more substances than any other liquid, it is a polar molecule, has +ve and -ve charged regions, can disrupt forces that bind other molecules together (Cox, 2016).

Stretching my capacity to understand, with my elementary knowledge of chemistry, I swim and puzzle, swim and puzzle, wondering my way into this new imaginary of water as the beginning of all life on Earth. I grab a pen and notebook and scribble notes to revisit later in this world-becoming with water:

> Our blue planet is an interconnected matrix of rivers and oceans transporting the dissolved ingredients of the Earth integral to life. Salts and nutrients essential to life are also transported through our bodies in water, the intimate connection every living thing shares with our blue planet, water delivers the ingredients of life. Hydrogen is the simplest chemical element, its atoms consist of a single proton called the atomic nucleus surrounded by a single electron. Oxygen is a nucleus of 8 protons, 8 neutrons plus 8 electrons.
>
> (ibid.)

Swimming again, I am losing and finding my way, losing and finding my way in the dark of limestone caves deep under the world's oceans where fresh water from within the Earth's core eats through solid rock and seeps through fissures to mix with salt water. Creating a boundary where dense and salty sea water sinks through less dense fresh water, creating a beautiful phenomenon known as halocline. 'A beautiful phenomenon of a halocline', search for dictionary to understand its meaning: 'In oceanography, a halocline (from Greek hals, halo- 'salt' and klinein 'to slope') is a subtype of chemocline caused by a strong, vertical salinity gradient within a body of water.' Still under the sea, swimming and swimming, there has never been a place to find the first mixing of salt and fresh.

And then comes the moment I'm waiting for, standing on a rock at the edge of the ocean in Iceland, a hydrothermal vent has been found with an outpouring of fresh water heated by Iceland's geothermal energy. Demonstrating with a battery, the merging of salt and fresh water creates an electric current, the same effect produced by the combination of oxygen and hydrogen atoms in that very first moment of life on Earth. The very simple hydrogen atom joins irrevocably with the more complex oxygen atom to form water, H_2O, the substance of life. Fresh water taken from the vent is alkaline and the sea water is relatively more acidic, an excess of protons on one side and there's that cascade of protons, the same thing that is happening in the cells of the human body, and of every living thing on the planet, the cascade of protons.

I return to the cascade of water in my swim hole at every day's beginning of the world with water. Swimming in the cascade of protons, the beginning of life on Earth, the elemental theatre of life, the fresh water that merged with the salt, the moment when the inanimate became the animate. Human always already more-than-human, our most distant ancestor not a living thing at all, but a geological thing, most likely a vent in the ocean 4 billion years ago. Swimming, gliding, with the water's flow in my swim hole, cascade of water over rocks pulls body in its inevitable flow, tumbling hundreds of kilometres back down the ages to my one-time island home where salt water meets fresh in Broken Bay. What does the meeting of fresh and salt water produce on Scotland Island, the birthing place of four babies?

Becoming with water: a journey of watery memories

Becoming with water invites a memory visit to Scotland Island to dwell in this space where fresh water and salt water meet. It is here that I enter the *entre deux* of broken time, the time before that cannot be reconciled with the time after, in the dramatic shift from a tiny island in Broken Bay to a remote desert community. Broken Bay is a place of broken memories, severed at that moment of watching the materials of our daily lives disappear on a barge, not to be seen again for 6 months. The time to write on Scotland Island is about remaking memory traces in body and mind, reliving the watery beginnings of family lives. A time to linger in multiple materialities, the materiality of bodies, of places, of landscapes, of objects, all carried across time and space in the waters of everyday life.

A water memory journey

I have travelled so many times with the water's flow from where I swim now to the place near where the Hawkesbury Nepean River ends its journey in the great salt water Pacific Ocean. I have followed it on Google maps, tracing the long thin blue line of fresh water hundreds of kilometres from the Nepean River at Emu Green, arriving at a wide bay broken by small islands, the place where salt water from the ocean mingles with the fresh of the river, Broken Bay. Every time I arrive at one small round island, *Scotland Island*, but I have never returned there since the great leaving so many years ago.

It is time to make this memory journey. Desiring to rematerialise the life of those early watery beginnings, I book a homestay in an artist's residence. It is the only place to stay, on the opposite side of the island to where I lived so long ago. I leave the place of my river and follow the line of the coast past the northern beaches with their waves crashing in from the vast Pacific Ocean.

Turning inland the road passes through tall rainforest until it opens out to a tiny jetty at our designated meeting place. Our host greets us in a tinny with small outboard motor and we climb in to traverse the bumpy waves to get to our destination.

Choppy water memory

Body grows tense against choppy waves, overwhelmed with sudden unwelcome memories of that first home coming. Bumping across white crested choppy waters, watery tears rolling down face, arms cradling first-born precious baby girl, tightly held against rocking of waves. How can we live here in this place of waters where the only way off the island is a clinker-built rowing boat, navigating shark-filled waters between an island home and mainland? It's the middle of a sunny day but dark waters flinging us to and fro are a portent of darker undercurrents of danger. Rock and bump, rock and bump, cradle new baby ever more tightly against my warm body. I'm jerked back into our current reality as we land at a jetty, tie up the boat, and climb the steep winding steps to our new home.

Water and light memories

Memory journey enters a temporary lull as, released from body memories, we arrive at a jetty, the landing place of our temporary home, an artist house of water and light (Figure 3.1). Here in this house of art in the aura of paintings of light on water, becoming with watery memories is all about the play of light and water reflections. Everywhere in the house we sit facing images of the changing light in Paul's paintings. The earthy colours of Hawkesbury's lithic sandstone shimmer with light on water, reflecting the movement of water's endless flow from where I live now to where I lived then, with my babies, on the other side of this island world. In another, the play of trees, water and light creates watery memory reflections from long ago times.

First light, 5.21 a.m.: dawn light

First waking on island, where will light come from across this body of water, east is direction of sea, the way we face here now is back against the flow of the river. Body memory tells me I am on the wrong side, the other side of the world. Surrounded by water. I think of the light, the way the light comes in the morning, the great turning of the world's circle and how from my bed on my river I face east, the direction of the sun coming up over the river valley, over the horizon to where the sea lies, the vast Pacific Ocean. And in our little long ago house on the south side, the opposite side of this island, the light was totally different, memories of light and dark, seeking sunlight from the darkness of trees. Light is memory's guide.

Figure 3.1 Water and light

Memory walk 1, 5.51 p.m.: lost in dark night

Tonight walking in the half-light I am lost, lost, lost. Walking, walking, walking, I cannot find the way back to my temporary home, trees have grown monstrous, looming dark over road, houses are larger, grander, lights shine from them, but they are not the place where we are staying. There are no street lights, no house numbers, and the road which makes its way up and down and in and out turns out to be much longer than its 3 kilometres around, the road goes on and on. Lost in the dark.

Memories surface of searching for light in our tiny house. Tall leadlight casement windows open out onto the dark of trees, bringing dappled light inside, casting reflections on newly painted walls and the paving of back deck, a nest to shelter a new baby. I remember the constant search for the warmth of sunlight, for being in the sun, sunning my babies, bare skin, bare bodies in the shafts of sunlight that penetrate the bush on a winter's afternoon. Picnics on soft bright green grass in the front yard in the tiny patch of sunlight streaming through the dark of trees. Daily walks to the little beach beside the wharf, to swim and sun in the warmer days, days of sunlight, salt water, and joy.

Finding home and a new baby

Walking on, searching for the reassurance of home, I come to a place in the road that feels like the right terrain, a huge multi-storey house beside a tiny inconspicuous house painted dark green, is that our temporary home? It could even be the right house but unsure in the dark, even of the way in, I keep walking on, eventually turning back, back the way I have come, a backward memory journey cascading down the years of time.

As the dark thickens in the disorientation of return, dwelling in two parallel universes, another body memory of becoming with water surfaces. I am 20 and in labour with my second child. It's midnight and dark envelopes as we walk to the wharf where our boat is moored. I am labouring and my husband is carrying our baby girl, as we make our way in the dark to our boat, wade through knee-deep water to climb in. I row across choppy salt waters so he can hold our baby girl. At Church Point, we leave the boat and wade ashore, jump into our car, drive to Mona Vale hospital and soon give birth in a rush of waters. A beautiful baby boy slips from his watery world to take his first precious breath of air, that moment of transition from water-dwelling being to earth-dwelling creature, the transition that marks the beginning of a new human life.

I stride on confidently remembering a body strong and fit from island living, a welcome and easy homecoming. Two babies now in a tiny fisherman's shack, pretty leadlight window's play of coloured light. Tucking small bodies into soft calico homemade crib, and fabric-lined cot, to tinkle of windmill music box as I find my way back to our artist home of light and water's memory reflections.

Memory walk 2, 9 a.m.: sun of everyday life

Walking into bright sunny morning, sun warms back, mother and two girls load bits of rubbish, plywood, and pipe, onto moped. There was never any garbage collection on our island. 'I'll take you to your workshop now', the mother says and off they tootle. I wander off in search of my house on the other side of the island world, wondering about Deleuze and the problem of memory, memories are fixed, not rhizomatic, he says. My memories are not fixed, they are watery memories, mobile, either flowing deep in my body and inaccessible to the conscious mind, or displaced, blurry, to non-existent. I am tired and the effort of memory is great. I walk and walk and walk, taking photos of vaguely relevant things as random advertisements bring to life the everyday present of living on an island surrounded by water:

> The firefighters of Scotland Island are cooking a SUMPTUOUS SPIT ROAST, dance to Rachael's acoustic project; with a fat pink cartoon pig over pink flames on vivid blue background …
>
> THE ISLAND KINDY is now enrolling! Long day care, half day care, after school care; with two smiling children holding up the sign.

> KINDY CAFÉ AND SECOND HAND MARKET: let's save landfill and give quality goods a good second chance; with colourful cartoon characters, clothes rack, table with goods and chattels, seagull and dog.

Body memories of daily life come flooding back as I walk. A tiny fishing shack visited before we lived there, low on the south side of the island, not far from ours, built over water lap lapping underneath. A house of babies, a milky space of mothers breastfeeding, a fluid place of water and babies, a foretaste of what was to come in my own life. Starting a playgroup, gathering with other young mums and their babies in the old fire station. Every month walking with the babies around the island to the other side, meeting other mothers in a monthly salvation from isolation. Walking stories, the grand old house of the queen of the underworld, Tilly Devine and her infamous life of crime and illicit wealth. Poet Rudy Krausman's house, designed and built by a Japanese architect from island timbers. A poet's house of yin and yang, light and dark, a place to write poetry of island life. Stories of an island place of myths and outlaws, a community of writers, artists, musicians, actors in the broken watery isolation of Broken Bay.

Memory walk 3, 12 noon: finding home

At noon, I set out again, still searching for my island home, the material remnant that holds the traces of that watery life. Walking, sensing, unsure where I am, even when I walk past a place with a strange familiar body sensation, not sure, keep walking. As the dirt road curves back around towards the other side I know I am lost again and have gone too far. I walk and walk, always with the blue of water alongside, through trees, bushes, houses, water guiding me. A wharf with a sign, *CARGO WHARF*, brings a surge of relief of recognition, the place where our worldly goods were loaded onto a barge, a liminal place that marked the end of island life.

Walking on, there are no more memories for me here, so much an outsider, so lost, until body senses suddenly change again. Another wharf, this time not for cargo, but for people, a jetty stretching out beyond low tide. A ferry stop, arrival and departure place of people crossing the waters back and forth to the island from school, shopping, working, visiting, all manner of things. I walk down the steep steps to *Bell's Wharf*, onto the jetty with its small timber shelter at the end jutting out into the blue. All the body memories flood back, the steepness of the steps, the familiarity of the ferry run, the people getting off and on, the comings and goings of daily life on an island.

At the bottom of the steps beside the wharf is a small strip of sandy beach, present and absent with the tides, growing and shrinking with the ebbing of the moon's waxing and waning. It is crowded now with houses and jetties, a jetty every few metres, but it is the same sand, the same beach, the same lap lapping waves as one day of a memory that remains fixed, indelible, surfaces from the depths. One baby playing in the water, the other, a tiny one on the sand. The water baby starts to sink, up and down, up and down. I run into water to rescue, panicked to leave smaller baby behind but I cannot, cannot let the water claim back that one precious life. Pulled back from water's strange desires, we are all back on the beach again warming in the sun. There are many days of sun and joy, but it is the memory of the power of water that remains.

Fulfilled with the memory traces of everyday life, of walking up the steep steps carrying shopping, babies, all sorts of goods to be carried, no cars then, I know the way home. At the top of the steps, body turns right, walking along the dirt road, knowing the familiar terrain, body sensing how far it would be. The road veers upwards on a steep incline then down into a gully. Then as it evens out, the sun comes through and it is the same place that I passed before, feeling unsure but with a sensory feeling of recognition. This time I know I have finally arrived at the place of my island home. A large two-storey house sits against high trees and bush at the back, our tiny fishing shack gone. But in front, bright green grass with shafts of sunlight, a place to picnic with babies in the warmth of sun in the middle of a dark winter's day. It is my place, I am home.

Memory Walk 4, 6.45 a.m.: stillness of water dreaming memory

Back at our homestay, I hop into the water for a quick swim in cold fresh water flowing in from the ocean. Out of the water, body warms in sun of artist's boathouse made of old timbers rescued from their salt water home. An artwork of excrescences of decaying timber and metal's melding with salt water spray, producing material artforms counterpoised to delicate play of water and light. Tired from all this walking and memory work, body sinks into warmth of sunny timber bench to sleep and dream in rhythmical lap lapping of Scotland Island waters. Dreaming, of a belly growing round again, another baby, making another watery baby home. It's the dark of midnight when tight extended belly begins draining of fluid, no pain just eternal seeping. The waters continue to seep out, draining away the watery life fluid home of the babies living within.

Tremor of fear courses through body for twin babies growing inside. Clamber labouring down dirt track to rowboat for third time in less than three years. Lap, lapping of water, pull of oars, as boat slides through dark waters of Broken Bay. Broken, the waters are broken. Lying flat on hard hospital bed, emergency ward, waiting hour upon hour upon hour upon hour until morning dawns and twin boys slip unknown from my body, abandoned by their watery world. Half-waking, half-sleeping I stroke excrescences of salt on sides of timber bench with soft fingers, longing for feeling of tiny not yet formed beings, forever lost to mine. Waking is a strange relief, not that I am no longer in this dream, but for the gift of this water dreaming memory's return, and I think of the dolls.

The dolls

It might have been after this great loss that the frenzy of dollmaking started. I remember sewing doll's bodies from calico, a slightly imperfect weave of unbleached cotton, stitched into body parts and filled with white fluffy fibre. All of the legs, all of the arms, and all of the bodies, hanging on a line suspended on kitchen room ceiling, above the babies playing underneath. After all the dolls' body parts were made, the fun part begins, the time to add colour and style to the doll babies just as the sunlight adds colour to our world. Multiple colours to choose in fine Laura Ashley cotton for clothes, thick strands of carpet wool for hair, coloured felt for shoes, and embroidery cottons for faces. Harmonising shades of a single colour for each doll, one in delicate shades of blues and greens, like the colours of the water lapping around me now.

Closing the curtains of watery baby memories

The lapping of water was always with us in our island home. Closing the curtains of night for babies sleeping in a home-made crib, stitched from the same soft calico as the dolls, and held suspended between two crossed timbers at each end. An antique oval timber table tilts on its side to make more space in the tiny shack, a chaise longue of carved mahogany with woven tapestry is placed underneath the pretty stained-glass windows looking out into the ever encroaching bush. The lap lap of water in the Broken Bay lulls us all to sleep and dream in our island home where the waters of migration awaken at the dawning of every new day.

Waters of migration

Scotland Island, 5.21 a.m.: dawn of new day

It's a pearlescent morning as if a sky painter used the same opalescent paint as the artist, soft grey with lights shining through cloud. The light of day comes in blankets of soft dark pink cloud stretched out across the dawn sky, over a soft mattress of deep purple and highlighting a dark line of trees on the other side of the broken bay. The waters rippling towards the shoreline are lit pink and moving in the light. Seconds later it's gone as the sun rises high in the sky, this water memory journey is all about light and calls up how the light shines on the other side of the world.

Every day waking in the artist's house on the island, body memory tells me I am on the wrong side, the other side of the world, surrounded by water. I think of the way the light comes in the morning, the great turning of the world's circle. Wondering how it was for my grandparents who migrated from Scotland to Australia, from the other side of the world, across the endless Pacific Ocean, that lies to the east of where I am now, merging its salt water into the fresh at Broken Bay. And the even more profound *entre deux*, the *entre deux* of the migrant where the world before does not mesh with the new world to come. I wonder about what they brought with them and what they left behind. My father's mother came from Scotland in the hull of a boat looking after other mothers' babies and then returned to Scotland, alone, to give birth to her own. How old was she and why did she make that incredibly long and arduous journey to the other side of the world to give birth to her first born, alone? Scotland was always home to my four grandparents. Scotland, on the other side of the world.

Waters of migration: material connections

We know little about the circumstances of my grandparents leaving Scotland. It was a time of dispossession when the English bought up large tracts of Scottish lands and Scots people became poor city dwellers crowded into unsavoury tenement buildings. Despite the long and arduous journey, and the circumstances of their leaving, even after they had settled in Australia, they all still thought, dreamed and spoke of Scotland as home, in the forever longing of the displaced migrant. My grandparents spoke with a Scottish accent all of their lives, my maternal grandmother spoke a strange language that seems to be known as Doric and I can remember her saying that I had *a sair face with a ne clute on it.* I always knew what these sayings meant but never any exact translation.

Scottishness was celebrated on many occasions. Every year my parents and grandparents attended the annual Drummoyne Scottish Ball, wearing kilts in the tartan of the Stewart clan but we children didn't know what happened there. As children, we were taken on celebratory picnics to Parsley Bay on Sydney Harbour with the Drummoyne Scottish Society of the Scottish diaspora. Scottish kilts were worn, my grandfather played the bagpipes, and there was much Scottish singing and dancing. I remember my Aunty showing me postcards from Scotland and describing the place in intimate embodied detail but as far as I know she had never been there, embodied Scottish memories had defied the waters of the long journey between home and the new/old country.

It was not until I travelled to Scotland after my four children had left home that I re-discovered a sense of connection. Scotland immediately felt like home, my fair Scottish complexion belonging in the gentler soft atmosphere. Tears welled up walking across in Scottish highlands at the sound of the once-hated bagpipes resonating across the moors from the valley far below. A visit to a tenement house maintained in its original state brought waves of spatial and material recognition, the furniture, the crockery, the arrangement of the spaces, the tucked-in bed with its curtains closing off the world, all felt like coming home.

Some of it I recognised from the crockery that once decorated the kitchen dresser, that we bought for our island home: a cream china chook, hollow inside to house the eggs that once sat in my maternal grandmother's house; an ornate china butter dish complete with curlicues and muted image of wild blackberry flowers, fruit on top and leaves in mauves and pinks on its sides. A small fine jug, golden rose on one side and green and gold butterfly on the other. A dark brown vase with some figures emerging from the dark as if lit by a lamp, memories of my grandfather's job of lighting the gas lights in the dark nights of Scotland, only a boy, before they came to the other side of the world. These things were with us on Scotland Island, a tangible connection to the endless travel across the vast Pacific Ocean to the other side of the world for a new beginning mediated by water.

Rejection of my Scottish heritage, the sense of being forever migrant, led to a lifelong search for belonging in the Australian landscape through deep relations with Aboriginal people, communities, Country, and land, beginning with the remote northern territory, the first big in-country migration from my island home, and continuing throughout our island continent.

Waters of first contact

Here in this house of water's light and reflections I pore over history books and piece together fragments of the story of the beginning of Scotland Island. I learn that the waters surrounding our island home were formed as drowned river valleys, created when sea levels rose during the last ice age. Millennia later, on Sunday, 3 March 1788, only five months after the arrival of the first fleet, Governor Phillip and his party left Port Jackson to explore Broken Bay by cutter and longboat, water being the only means of travel in the new colony. Despite his greeting on arrival by 'one man and five women in three canoes', Phillip did not hesitate in naming 'the finest piece of water I ever saw', *Pittwater*, for the British Prime Minister, William Pitt the Younger.

In these historical readings there is so much sense of loss, so much missing. I long to become with those first meetings between the new settlers and Aboriginal people but the early accounts are few, dates are confusing, and they are recorded only by white invaders, so they present a very limited understanding of these encounters.

I dwell in the little I can find, in the materiality of Aboriginal livelihoods: bark shelters, nets, fishing lines, spears and a stone hatchet, water-carrying implements, and in the incorporation of new materials like a piece of coarse linen observed to have come from exchanges with the settlers themselves. These tiny materialities offer something small and tangible to my desire for embodied knowing of the ways of life in this place of the meeting of the fresh and salt waters where there was so much to be gained and lost.

Hawkesbury River waters

There are many accounts of the very early exploration of the Hawkesbury River which was reached by sea from the site of the colony at Sydney Cove. After they had been met by the boats of local Aboriginal peoples in the first expedition of 1788, the party went on to explore the arms of Broken Bay. They are described as 'sighting' the mighty river the Aborigines called the *Deerubin*, whereupon they rowed some 20 miles before returning due to insufficient provisions. The explorations continued in 1789, in search of suitable lands for settlement and agriculture, and resources such as timber, with which to construct the growing colony. They travelled as far as the confluence of the Grose and Nepean Rivers at Yarramundi where the party found that there was too little water for the boats to advance any further, describing the marks of the vast torrent that must pour down from the mountains after heavy rains. Some little of the sense of the place is retained in these water memory traces of the early accounts:

> The low grounds at such times, are entirely covered, and the trees with which they are overgrown, are laid down (with their tops pointing down the river,) as much as I ever saw a field of corn after a storm, and where any of these trees have been strong enough to resist in any degree the strength of the torrent (for they are all less or more bent downwards) we saw in the clefts of the branches of such trees vast quantities of large logs which had been harried down by the force of the waters and lodged from thirty to forty feet above the common level of the river; and at that height there were great quantities of grass reeds and other such weeds as are washed from the banks of the river, hanging to the branches.
>
> (Hunter, 1789)

Nepean waters and Boorooberongal arrival in settler imaginaries

In April 1791, Governor Phillip undertook a further expedition to determine if the Hawkesbury and the Nepean were the same rivers. This time, substantial interactions with Aboriginal people were recorded and inscribed on plaques at the site of Yarramundi today. It tells how they encountered the Boorooberongal clan (of the Darug peoples), whom the Aboriginal guides referred to as 'climbers of trees'. One of the clan demonstrated tree climbing to the exploration party, bolting up a tree by means of toeholds or notches that had been cut into the trunk. Relations between the party and the Aboriginals encountered were cordial and three of the clan joined the group for a period – Gomberee, his son Yellowmundee (Yarramundee), and grandson Djimba (Tench, 1791).

There are many accounts of close and enduring relations between Aboriginal people and the English officers who hoped the narratives they were living and writing would have the happy ending of a peaceful co-existence, over the first weeks and then during exploratory journeys to Broken Bay, they described many scenes of frolicking and fun, games, laughter, jokes and 'dancing with strangers' on those beaches (Karskens, 2010, 49). A close analysis of one of the paintings that recorded these happy times reveals the tensions that lay beneath them:

> Some of Lieutenant William Bradley's watercolours depict moments of major breakthroughs in early race relations: the first meeting of Eora women dancing together at Broken Bay … on the southern shores of Broken Bay where the series of headlands stretch down towards the crouched outline of Lion Island the dancers holding hands are men, for the women, as always, stay out of the water in their nowie watching. The longboat and schooner are close to shore, too, manned and ready for a quick getaway, the rowers resting on their oars. Underneath the surface of

> innocence and play, this is an edgy picture, reflecting the tensions and watchfulness of these first encounters. And onshore, right at the centre of this happy group are two figures who are not dancing: they are wearing red coats and have muskets resting on their shoulders … there could be no dancing, no rambling explorations, no meetings at all without guns and redcoats.
>
> (ibid., 50)

Piecing together the many fragments that emerge from these accounts, and reading them diffractively against each other, produces a sense of the swirling unpredictable chaos and horror emerging in this meeting place of fresh and salt waters.

Life and death on the waters: gal-gal-la

One of the most horrendous outcomes of first contact was the introduction of diseases that arrived with the new settlers via the contact of the waterways. By the time of the exploratory journey of 1789, just one year after the arrival of the first fleet, it is reported that *gal-gal-la* (smallpox) had struck. Aboriginal people, already infected with venereal disease, succumbed to this new infection, and by May 1789 the settlers discovered more and more bodies of Aboriginal people in and around the coves and inlets of Sydney harbour. It is now believed that over 50 per cent of the population of south-eastern Australia perished within three years of the European arrival.

One of the saddest fragments of all is an image of a scene related to the smallpox epidemic in a drawing reproduced in these historic documents. The drawing portrays an Aboriginal woman crouching naked in a bush shelter facing away from a party of five men dressed in military garb all looking at her as she folds her body over a small baby lying face down on the ground beside her. Her head is turned back to look over her shoulder in the direction of the men. The title on the drawing says 'A frightened Aboriginal woman, recovering from smallpox, encounters Governor Phillip's exploratory party in June 1789, as depicted by John Hunter'.

I am horrified at the encounter portrayed in this image. I imagine myself in the position of a young mother, cradling my new baby, terrified at the spectres of death that have suddenly come upon our lives, newly recovered, and barely able to care for my baby. I am confronted by the group of men in strange red and black clothes who all stand and stare at me, frightened for my life and my baby traumatised by illness and suffering I have nothing to cover and protect me. I turn to keep an eye on them sheltering my baby with my body as the only protection I can offer.

My horror is not ameliorated but made more complex when I read later that

> On their outward journey in 1789 the party had found a terrified young woman, so weak after surviving smallpox she had not been able to run from them. They made her comfortable, brought her food and water, lit a

> fire, spoke the soothing words they had learned from Arabanoo. On their return to Broken Bay they found her again, much recovered, with her husband and child.
>
> (ibid., 378)

Despite these pleasantries, there continued to be overwhelming loss in the life and death struggle in this place where

> the rivers and the lagoons were rich in fish, birds and edible plants that provided the Aborigines with most of their foods, their campsites and their great gathering places, the river banks were seen to be churned up by the digging sticks of the women harvesting the nutritious tuber, *dioscoria transversa*, and the rivers also provided starchy edible tubers and floating nardoo fern (*Marsilea*) besides fish, freshwater mussels and turtles; donkey orchids and chocolate lilies grew in the river flat forests.
>
> (ibid., 463)

Plundering the land: shells, shingles, timber, salt and fish

Just as the waters provided the new settlers with the means of transport in the new colony, they also enabled them to set about harvesting whatever they could take to supply their needs for food, building materials, transport and more. Aboriginal shell middens, sometimes as high as 10 metres were plundered, to provide lime to make mortar for buildings. Shells from Broken Bay and Pittwater were offloaded at Lime Burners wharf where they were burned until the 1850s and 1860s. Traces of these shells can still be seen in the historic buildings around Sydney today.

Casuarinas were prolific around Broken Bay, Pittwater and the Hawkesbury, as they continue to populate my river today. In those early days of settlement, shingle cutters set up camps to fell and split the tree trunks to make shingles for settler housing. Another species of casuarina was used in colonial furniture, and the red or forest oak species was felled to fuel bakers' ovens as the wood burned quickly.

Loads of snapper and bream were dried cured and shipped in barrels, reaching as many as 200 barrels in one season. On Scotland Island, Andrew Thomson harvested 90 kgs of salt per week, dried in an oil burner, to service the growing colony.

The waters provided all of this to the new settlers, the very materiality of the river harvested relentlessly to meet the needs of the newcomers, somehow learning and not learning how to live in this old/new land.

Naming and claiming

From the very beginning of first contact, in the meeting of the fresh and salt waters the invaders began to name and claim the land as their own. In 1788, they named the small island that sat in the middle of Broken Bay, *Pitt Island*, after the same British prime minister of the time. Later, when the land on the island was 'given' as a land grant to a Scottish settler, Andrew Thomson, he renamed the island *Scotland Island* after his homeland.

The land that had once provided everything for the peoples in whom its care was deeply entangled were dispossessed and land wars eventually broke out, brutally leading to murder and mayhem in the 1790s. These stories sit alongside other stories of lifelong friendships, extraordinary loyalty, generosity and devotion. Just one of these stories is of Bungaree, who left his home in Broken Bay in 1799 and sailed to Moreton Bay to lead yet another exploration party. Bungaree was the first Australian to circumnavigate the continent as all the rest of the exploration parties were European. He was also the first of a long line of largely uncelebrated Aboriginal explorers who went into unknown country, guided the parties, found them food and water, and saved their lives by acting as peacemakers and emissaries (ibid., 428).

Despite all of my previous theorising of the contact zone where 'peoples geographically and historically separated come into contact with each other' as zones of 'possibilities and perils' (Pratt, 1991); a space of 'radical openness, a margin, a profound edge' (hooks, 1990, 149); 'places of profound emotional pain, work that the soul performs' (Anzaldúa, 1987, 79); 'a space of painful and temporally situated coalition work' (Haig-Brown 2001, 31); and 'a discomfort zone of productive tension' (Somerville and Perkins, 2005, 265), none of these words comes close to describing the affective depth of response to the stories of encounters in the broken waters of *Broken Bay* as I turn to the paradoxical story of the watery beginnings that enabled the growth of Sydney into the mega city that it is today.

Water's beginnings of the mega city of Sydney

The story of Sydney's growth is the story of water, the availability of the first essential ingredient for life. The first water supply in the newly forming colony was what is now known as the Tank Stream, the reason for the location of the colony. Once a thriving complex ecosystem, the Tank Stream was where estuarine, rock platforms, creek, open forest, and wetland came together to offer a rich diversity of food and materials for Gadigal people's everyday lives.

The intermingling of people and land changed very soon after the new settlers arrived. Tanks were cut into the bedrock of the stream, soon after were enclosed with a fence in an attempt to prevent stock muddying the water. A short time later trenches were cut alongside the stream in an attempt to catch runoff before it could enter the stream, but due to the increasing population, number of buildings, and the loss of trees as land was cleared, runoff increased. Human and animal wastes, and domestic products were discharged into the stream which became an unofficial sewer by 1826 and an official sewer in 1857. The Tank Stream could not meet the needs of the growing colony, increasingly polluted by runoff from the settlement and it was finally abandoned in 1826.

The first major intervention in my Nepean River was the upper canal system, constructed between 1880 and 1888. The scheme involved tapping the headwaters of the Nepean River and its tributaries, the Cataract, Cordeaux, and Avon Rivers. A convict-built system of tunnels, aqueducts, and open canals diverted water which previously would have flowed down the lower stretches of the Nepean River to Camden and Penrith into the Upper Canal. From there, it was conveyed to Prospect Reservoir and thence to Sydney entirely by the gravity of water's flow.

Water shortages, Gundungurra, and Warragamba Dam

One of the first places in the Gundungurra traditional homelands that most appealed to the new settlers were the river flats of the Burragorang Valley on the meandering tributaries of the Nepean River. The settlers squatted on blocks using Aboriginal pathways to move through country, and although the Gundungurra resisted the taking of their lands, their claims failed. The narrow gorge at the exit to the Burragorang Valley was first suggested as an ideal dam site in the mid-1800s, but a century passed before construction techniques were advanced enough to build Warragamba Dam. During this time Sydney underwent a series of water shortages with four dams built in the Upper Nepean Scheme and a fifth dam at Woronora, but still water shortages continued to plague the growing city. In 1942, the drilling of rock core samples began in order to find the best site. Final approval for the dam design was given in 1946, followed by tree clearing in the Burragorang Valley and the diversion of the river around the construction site.

When completed in 1960, the 137-metre-high Warragamba Dam wall created a lake 52 kilometres long, drawing water from a catchment of 9,051 square kilometres, quadrupling the previous total storage capacity of all of Sydney's existing dams. The dam was the last, and by far the biggest, of the concrete dams built during the twentieth century. When completed in 1960, it was the world's tallest and largest domestic water supply dam to be built on stone foundations, the pinnacle in engineering achievement for this type of mass concrete dam building. Today it supplies water to 500 million people.

Closing the circle of ancestral connections

The most compelling and puzzling migration story of my ancestors was the story of the mystery child we knew as Wee Davey. Wee Davey was somehow connected to the strange story of my Nanna's journey back to Scotland to give birth to her first child, Nancy. She discovered in the process that her husband, David, had left a child behind in Scotland, another David who became Wee Davey to us. The story, as I remember it, is that Wee Davey was left behind in Scotland but seemed to appear and disappear again in family memories. Recently we discovered through his granddaughter Tracy Manatello, that he travelled to Australia in the same ship as my grandmother and her sister, enduring the same endless journey over the seas as a small child.

Searching for further details of her grandfather's early life Tracy was contacted by my sister through Wikitree:

> 16 September 2018
>
> Dear Jen and Margaret,
>
> Well, fancy that. Life is amazing, isn't it?
>
> Pop lived at Warragamba, and worked on the dam. Both my grandfathers and my Dad worked on the dam as well. We all went to Warragamba school and the dam was a large part of my early childhood – a lot of my cousins still live in the area. Mum and Dad are currently in SA 'grey nomading' so I will get in touch and see what I can find out for you.
>
> Take care
>
> Tracy Mantellato

> Dear Tracy,
>
> I'm still reeling from the shock of this revelation about Wee Davey and Warragamba Dam, the Dam on the Nepean River in the Burragorang Valley that has enabled the growth of the city of Sydney into the mega city it is today. I've been fascinated by Wee Davey for as long as I can remember, as the mystery child in our family, and have equally been fascinated by rivers/waterways for as long as I can remember (Tarban Creek, Lane Cove River, Booralong Creek, Gwydir River, Blue Hole Lagoon, the Gwydir, Darling, and Murray) and now I am writing a book about walking our waterways as places of becoming (human differently). The third chapter is about Scotland Island as one of the first ever land grants in Australia because all travel was initially by water, and of course Scotland Island is named after Scotland as the homeland of the second grant holder. The Scotland Island chapter is about beginnings with water, from life on earth, to the birth of my babies, the very early explorations of the Hawkesbury Nepean River system and its Aboriginal custodians, and the arrival of our family from Scotland by the waters of the Pacific Ocean.
>
> Thank you,
>
> Margaret

> Dear Margaret,
>
> This is a photo of Wee Davey, Pop, David Jnr, in diving bell (Figure 3.2). Working on construction of Warragamba Dam.

> Dear Tracy,
>
> I dwell in the images of Wee Davey in his diving bell, about to launch into the dangerous underwater work of constructing the largest dam of its type in the world. A strange sense of encirclement (rather than completion) emerges as I near the end of this story of beginnings with water. I circle

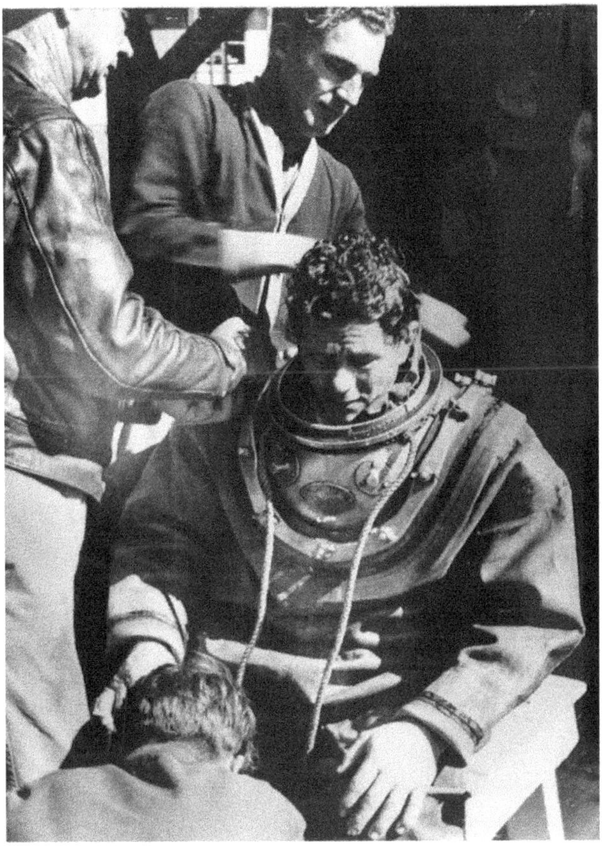

Figure 3.2 Wee Davey and son, Warragamba Dam construction

back to the hundreds of creeks, the nourishing veins that gave life to the new colony, and the Tank Stream that provided its fresh water, and their eventual loss. The waters of Broken Bay, where the first settlers encountered the peoples of this ancient land with both great brutality and great affection. The broken waters that heralded the arrival of my new babies on the island, named after our ancestral home. The mixing of the fresh and salt, as the very beginning of life on earth, somehow becoming a mode of representation for the surprises of the beginnings with water. Finally coming to the end of the Wee Davey mystery that connects my very DNA to the dam that provides 500 million people with about 1,700 million litres of water a day. The same dam that drowned the ancestral creation sites of the songlines of the Burragorang Valley, songlines that nevertheless live on in water's eternal flows through the bodies of the world.

Thank you,

Margaret

References

Anzaldúa, G. (1987). *Borderlands/La Frontera: The New Mestiza*. San Francisco: Spinsters/Aunt Lute.

Cox, B. (2016). The Moth and the Flame, Season 1 of *Forces of Nature*. BBC. Available at: www.bbc.co.uk/forces of nature

Haig-Brown, C. (2001). Continuing collaborative knowledge production: knowing when where, why and how. *Journal of Intercultural Studies*, 22(1), 19–32.

hooks, b. (1989). Choosing the margin as a space of radical openness. *Journal of Cinema and Media*, 36, 15–23.

Hunter, J. (1789). Plaque at Yarramundi Reserve.

Karskens, G. (2010). *The Colony: A History of Early Sydney*. Crows Nest, Australia: Allen and Unwin.

Pratt, M.L. (1991). Arts of the contact zone. *Profession*, January, 33–40.

Somerville, M. and Perkins, T. (2005). (Re)membering in the contact zone: Telling, and listening to, a massacre story. *Altitude*. Available at: www.api-network.com/cgi-bin/altitude21c/fly

Tench, J. (1791). Plaque at Yarramundi Reserve.

4 Songlines

Walking the songlines of the singing painting river

With Leanne Tobin and Jacinta Tobin

Introduction

Walking the songlines invites the reader into a becoming through walking in deep time, and through walking the contemporary songlines of the Darug peoples today, in answer to the challenge of the mighty Nile River, that we do not need another explanation for what we have lost, but an exploration of what we may find. Songlines are walking trails that follow the stories of the great ancestral beings as they created the world and all that is in it. The songlines that once echoed throughout the river's valley are now gone, as are the songlines of the Burragoorang Valley where they are now submerged under the water that feeds the city of Sydney. Songlines, however, are resilient, they live on, they are recreated in song, story, art and language. The Hawkesbury Nepean songlines are explored through two long and challenging walks along the rugged bushlands of the Nepean trail, walking trails for movement between places of creation. New songlines hold the contemporary history and cultural memories of singing the rivers in song and ceremony for Darug peoples today. Songlines invite the reader to meditate on the presences and absences of these river places with Darug artists, song writers and cultural knowledge holders.

The river is always there
Walking puzzling images and
pre-formed thoughts floating
like wibbles on river's still surface
where language and body meet
how to bring the river into presence is
the same as finding the way
to the swim hole
when track is hidden by long summer grasses
until I see slightest trace where each day
walking feet make their way to bliss of immersion
a walking trail that follows the songline to
a place of re-creation
where world is made anew.

Songlines: walk with me on a learning trail

In the beginning

A long time ago I was Nungarrayi, initiated into ceremony by Pintubi Luritja women in remote desert country, 246 kilometres south-east of Alice Springs in the Great Sandy Desert (Figure 4.1). The place where we sang and danced was a long way from the settlement. I had access to a four-wheel drive vehicle and was not employed by the government, being at home with my two oldest children, and pregnant with a new baby. This placed me in an ideal position to drive the women to ceremony, with the assumption that I would sing and dance as all women of childbearing age know ceremony. The drive was long, I had no idea where we were heading, there was no road, just driving through red desert country, the women gesturing and voicing to go here, go there, watch out for that one. I had no idea where we were, whether the petrol would last till we got back. It was a journey of profound trust and unknowing.

I have written about this many times but never used my 'skin' name, Nungarrayi. This kinship name placed me in a recognisable relationship to the community. It meant that my husband was Tjangala, my children Nampijtinpa and Tjampijtinpa. The feel of these names in my mouth still takes me back to the red dirt country and that place of ceremony. Most of all, my skin name gave the old Pintupi Luritja women, who had only recently been in contact

Figure 4.1 Pintupi Luritja women

with white society, a way to locate me in ceremony, a 'totemic' relationship to land and its living creatures. As a young woman of childbearing age they assumed I knew, so I learned by doing as they anointed my skin with kangaroo fat and delicately painted ochre designs on my white skin. They never referred to my pregnancy, except to lightly touch my growing belly with strongly approving smiles. I had some Pintubi Luritja language, but none of the secret women's language, learning was all through touch, touch of hands, skin on skin, feet on ground.

It is interesting to me now that it is a *skin* name. Skin was the mode of our communication. It was the pattern of designs on my skin in red and yellow ochres that held the meanings of my totemic connection to Country and its songlines. It was these designs that stayed with me when I returned to my house in the government settlement, day clothes concealing the designs, and the fatty, gritty feel of my skin retaining the traces of this incommensurable world.

I research my skin name now, searching 40 years later for the exact nature of the totemic relation to this country that has always eluded me. I have searched and searched. Many websites tell me about the complex kinship systems which I already know about from my direct experience of the old ladies coming into my house and searching my cupboards for food I could share with them; the people I was not supposed to communicate with; and others who would care for me and my children. But the question of the totem, the particular living entity (plant or animal) that was the core of my very being, continues to elude me, maybe protected and hidden because of its secret sacred nature, and rightly so. This information belongs in Country and in ceremony, always forever an inseparable part of that landscape and that moment of being.

So what remains is that there are no words of explanation, only body memories of gritty ochre paint on kangaroo fat skin, visceral chanting of voices rising up through red earth to deep within our bodies, stomping, hip swaying movement to rhythm of clapsticks. A liminal space opening up that never leaves me, a searching for songlines that sing the well-being of Country over a lifetime of collaboration with many Aboriginal peoples and their communities. Always and still the learner, understanding that I will never know the complexity of relationship with Country, but deeply committed to continuing the learning that is always so generously offered. Learning is slow and emplaced, gathering meaning in body memories, that accrete over time and place.

Mulgury as totemic relation

I learn from the Country of *Terewah*, the Narran lake, in far western New South Wales, from my very closest long-time collaborator, U'Alayi researcher and cultural knowledge holder, Immiboagurramilbun (Chrissiejoy Marshall). Chrissiejoy teaches me in paintings, oral storytelling, and translations from her Erinbinjori grandmother's knowledges passed down for hundreds of generations of her people. Her word for 'totem' is *Mulgury*, replacing the word totem as a crude translation from American First Nations' very different understanding. Through

Chrissiejoy's paintings, stories, and translation of her grandmother's teachings, I have something of a passage, a thread to the very being of the old ladies, in words that I never had at that time.

> In the beginning all was Mulgury. Only creative power and intent. Through the intent and power of our Creator, Mulgury reproduces into form to carve the beings and shapes of the world where the water meets the sky and earth sings the world to life. The pattern of life is Mulgury and Mulgury is traced in the Niddrie [the framework of the ancient laws within Niddeerie]. Every tracing, every rock, tree, plant, landform, the water, fish, reptile, bird, animal and Mudri is in the sacred relationship, through Niddeerie. The pattern, shape and form of Mulgury is life, and all is a continuing tracing of Mulgury.
>
> (Ticalarnabrewillaring, 1961, trans. Immiboagurramilbun, in Somerville, 2013)

I dwell in the idea of *Mulgury* from a place of unknowing and hope that through this dwelling and writing I will come to know. Immiboagurramilbun begins with swan. In the image that represents her ontology she paints four black swans. The first two swans are for her mother, and the second two represent the collective of water people, the Noongahburrah, her grandfather's people. In her culture, one swan on its own is a warning of bad omens. The swans are Mulgury, signalling their collective meaning as mythical creatures of the Niddeerie (dreamtime), as well as representing an individual's connection to a particular creature and its place. Immiboagurramilbun's mother is swan, Noongahburrah people collectively are swan. Swan belongs to the time and place of the creation of the land and people of Terewah, the home of the black swan, in the past, the present and the future. Those who carry that identity are both swan and place. Country, swan and person are together an ontological reality.

Immiboagurramilbun says that as an Aboriginal person you are given a Mulgury at birth and it comes with the responsibility for that animal or plant and is part of you and you of it. Part of that obligation is to learn all about your Mulgury and everything that is connected to it. If your Mulgury was the kangaroo, you would learn that you are related to the trees, the insects, the birds, the grass, the wind, the rain and all the things which occur and surround a kangaroo's life. You would spend years observing and learning about the life of your Mulgury – what it needs to survive and how it assists in the survival of other species. Most importantly, you would learn how all those things connect to yourself – how they all become your brothers and sisters, part of your family and about the responsibility which goes with that.

Contemporary songlines in Gumbaynggirr Country

In Gumbaynggirr country on the far north coast of New South Wales, still far from where I am living now on my river, I take a symbolic walk with Uncle Martin Ballangery. Together we retrace the memory steps of his regular summer walk from Bowraville in the hinterland to the river's estuary at Nambucca Heads stopping and stopping all along the way for stories.

> Here's the river, we walk from 'ere down, we walk all the way down there and just up a bit from there, Wirriimbi Island, there's another place. No, that's it – that's a property there, used to come across there, all the Old People used to meet there. On that side there. We used to meet 'em there. Down 'ere – probably because we'd meet 'em 'ere on this little part 'ere. On that little flat there. Here's a little creek goes up there, inside that there. And that's where all the women and kids use to wait for us as we'd walk down there spearin', and collectin' and gatherin' fruit.
>
> (Martin Ballangarry)

I learn that walking trails are dynamic interactions between people and place. Places always had their histories and the changing events that were significant in those histories are remembered in particular places as we walk. Everything signals other larger stories and connections, all is code for another series of stories, just like the epic Birrugan storyline, the creation hero of this Country. The storyline changes as the travel through country changes when further on down the walking trail we begin to map the water tracks that criss-cross the river on their arrival.

Songlines as walking trails of ancestral creators

It is on this symbolic walk, sitting with Uncle Martin, tape recorder between us, that I have the epiphany that the mysterious sacred songlines are actually walking trails. We have mapped the story of Birrugan, the creation ancestor, with other cultural knowledge holders, Uncle Ken and Gary, and surprisingly, it follows the exact same track that the annual summer walk to the coast followed. I learn that a songline is a walking trail that links story events in the path that the creation ancestors followed as they lived out their daily lives, walking through country, collecting food, living out the events that are marked forever in the landscape just as Gumbaynggirr people do, living their daily lives as a practice of Country, collecting prawns, crabs and fish from the estuary; turtles, swamp hens, eels from the swamp; pipis, googumbals and abalone from the beach and rockpools; mullet, tailor, mackerel, whiting, jewfish, bream, trevally from the sea; kangaroos, possum and porcupine from the surrounding bushlands; and turtles, eels and cobra from the river. In between all of these places, gathering native fruits such as lilli pillis, wild cherries, nyum nyums and pigface as they walked on their tracks through the dunes and coastal heathlands.

At each of these special places a song is sung and the songs are connected in the songline of the linking trails. It is along these linking trails that people walked to ceremony, making connections across Country and beyond. I learn that in Gumbaynggirr Country the stars guide the way on these linking trails, marked in the placement of stones on the high peaks, stones that follow the diamond constellation of the stars of the Southern Cross. The pattern of stars of the Southern Cross is also Birrugan, the creation hero whose actions created the walking trails, the line of story and of song. When language, story and Country are reconnected, storylines, the lines of the story, become songlines again, and Country can once again be sung into (well-)being.

Contemporary Darug songlines

I learn about Darug songlines in the Country of the riverlands when I meet Darug artist Leanne Tobin, and her sister, singer songwriter and language teacher, Jacinta Tobin. Leanne and Jacinta sing and paint the songlines of Darug Country. They teach me about Maria Lock, their great-great-great-grandmother who was Yarramundi's daughter. We meet at the Summer Institute of the Antipodes when we curate an exhibition, *Meeting of the Waters* to bring together local artists and artists from the Murray Darling Basin (Somerville and Power, 2016). Leanne exhibits her artworks and Jacinta welcomes the international audience to Darug Country in language and song. She offers a black feather of the crow, of special significance to Darug people as a gift to the conference. A year later Leanne, Jacinta and I performed and exhibited the songlines of contemporary Darug Country at the International Passenger Terminal, the place that was once Sydney Cove, the historic site of first contact:

> 'Connection to Country' is a term used by Indigenous Australians to describe their significant and complex cultural connection to landscape. 'Connection to Country' provides a model for ecology and sustainability in which human care for landscape is intrinsic and meaningful. Margaret Somerville, whose research investigates Indigenous understandings of Connection to Country, collaborates with Indigenous artist Leanne Tobin and her sister, singer song writer Jacinta Tobin, from the Darug language group. Together their work captures knowledge of country and continuous cultural connection to place in contemporary urban societies. Through Darug language, song, and visual images they will explore the conference theme of the third city through the network of rivers and their tributaries that form the veins of Western Sydney. They hope to open up a conversation about how 'Connection to Country' might inform Australian landscape architecture practice through the cross-cultural multimodal exchange enacted in their performance.
>
> (AILA, 2017)

We sang three songs and presented 73 artworks as a way of communicating to an international audience what Aboriginal understandings of Country might mean. It is these same songs and artworks that echo once again on the long walk from Emu Green to Yarramundee.

Leanne's painting

Leanne's painting of her great-great-great-grandmother Maria Lock now hangs on the wall of my living room, facing the nearby river with its water flowing to Yarramundee, the place where she was born, daughter of the Darug healer and new creation hero who mediated the time before and the time after everything changed forever (Figure 4.2). Every day I learn from this large, brightly coloured painting as it commands attention from its new home on the wall in my living room. The background is deep red ochre, the colour of song dance and ceremony. Sacred river gums hover over the landscape as shadowy spirit presences, almost concealing the presence of the Old People sheltered in their midst. An open wire fence cuts the landscape in half, impenetrable but failing to hide the Native Institution and its children on the hill beyond. A large totemic bird hovers overhead, and another smaller one perches on the ground. In the bottom corner the powerful figure of Maria Lock is writing with a feather quill, a possum, her totemic kin, draped protectively around her shoulder. Scattered beside her are the letters she has already sent, some transformed into strange black shadows, and others lying yellowed on the ground.

Figure 4.2 The Humble Petitioner, ochre, collage, charcoal and acrylic, 2000 x1000 mm
Source: Leanne Tobin, 2013

Maria Lock, a remarkable woman

Maria was a daughter of Yarramundi, the Boorooberongal elder who had met Governor Phillip on the banks of the Hawkesbury in 1791. She was educated at the Parramatta Native Institution, which she attended with Yarramundi's approval, from its opening in 1815. In 1819, aged 14 years old, she won first prize in the NSW Anniversary Schools Examination, ahead of 20 Aboriginal and 100 non-Aboriginal candidates. On 26 January 1824, Maria married carpenter Robert Lock, an illiterate convict assigned to Maria, the first official Aboriginal-British marriage in the colony. The couple were promised a grant of land and a cow. It required many years of struggle with the colonial administration, for Maria to get the promised land. Finally, in February 1833, she received a grant of 40 acres in the name of Robert Lock. In 1843, Maria returned to her ancestral lands at Blacktown, having inherited a 30-acre grant on the Richmond Road. Maria founded a dynasty that today has thousands of descendants living throughout Western Sydney and Australia (Tait, n.d.).

Yarramundi

'King' Yarramundi is also known as Yallahmiendi, Yallamunda, Yellomundi, Yel-lo-mun-dy, and Yellow Monday, in the variable spelling reflecting the transition from names as sounded in the landscape to the written word. He is described as a *Karidji* in Darug, translated awkwardly in English to 'clever man', or healer, because of the impossibility of finding words to communicate the depth of ancient knowledge and cultural practices held only by certain ordained people. Yarramundi is mostly known today for his role in brokering first contact with the Darug peoples and the new settler invaders. Yarramundi met Governor Phillip's second exploration party in 1791, searching to determine whether the Nepean and Hawkesbury Rivers were one. It was at the junction of the Grose River and the Hawkesbury Nepean waterway that the question was settled. Yellomundee Regional Park, named after Yarramundi, is located on the eastern escarpment of the Blue Mountains, the ridge of the Lapstone monocline that runs along the river on a long rugged walk from my river at Emu Green. Yarramundi's powers and his brokering role made him as a kind of new creation hero, heralding what was to come, and memorialised in Leanne and Jacinta's song:

> Yarramundi used to
> live at peace in this land
> around the Hawkesbury he roamed
> this old place he called home
> and his songlines are still hummin' on
> Yaarramundi, guess what
> the colours nearly gone a whole heap are dead
> and your family whitewed but your bloodline

is still livin' on.
Yaaaaaaarramundi.
They poisoned the river I'm sorry, polluted it may be
the times are all changed, I wish we could take
some responsibility cos Yarramundi
I never dreamt it would be this extreme
cos people are tryin' to tell me
I need insurance and diamond riiiiings
Yaaaaaaarramundi, yeah, we're hummin on and livin on and hummin on and livin' on
Yaaaaaaarramundi.
Cos people had smallpox but Yarramundi knew
he knew the ancient ways
the secrets aren't dead we've gotta find them in our head or Yarramundi please lead the way
did you understand about energy did you,
did you use your heart with your brain
you didn't need technology or complexity
and your pain, yes, it still remains
Yaaaaaaarramundi.
Yeah, we're hummin on and livin on and hummin on and livin on and hummin on
And *Yaaaaaaarramundi* I tried to walk in your shoes but your shoes
your shoes had too many thorns
cos the farmers have got your land now and your walking tracks no longer worn
but Yarramundi you used to live at peace in this land
around the Hawkesbury you roamed this old place you called home
and his bloodlines yeah we're still livin on *Yaaaaaaarramundi*
yeah, we're hummin on and livin on and hummin on and livin on and hummin on and
Yaaaaaaarramundi.

(Walk with the sound track, Jacinta Tobin, performed by Jacinta and Leanne Tobin, www.jacintatobin.com.au)

Walking with the rhythm of the song holds the traces of the land and its ceremonies, representing as Yarramundi does, the time in between the long past of Aboriginal presence in this ancient land, and the present time of white occupation. To sing along with the rhythm of the song is to incorporate Yarramundi's songline into one's walking body.

Wirrawee: Aboriginal women strong

To walk with *Wirrawee* is to walk to the rhythm of the women who carried the storyline into the future so we can walk with it today. Maria Lock, the matriarch, was the beginning of a long line of women who continue to keep the story strong today, her great-great-great-granddaughters Leanne and Jacinta Tobin. Leanne says she uses her art to tell stories and to connect the observer with the land they stand on, highlighting their shared role in caring for Country as well as conveying the intrinsic spiritual connection Aboriginal people have with the land (Tobin, 2017b). Jacinta describes herself a language teacher and musician songwriter who believes that music is a wonderful form of education and life. She works with schools, universities, community and government using her music and her community worker background to educate all on Darug language, songs and local Darug culture (ibid.). Leanne and Jacinta live, walk, sing, paint the songline of the singing painting river today, bringing it into the present to sing the well-being of Country.

> ***Wirrawawee***
> Wirrawee booboowhoo wirrawee booboowhoo
> yaaana nyiindi yaaana janawee
> Aboriginal women strong
> Aboriginal women strong
> Walk with you walk with me
> Janawee.
> Wirrawee booboowhoo wirrawee booboowhoo
> yaaana nyiindi yaaana janawee
> Wirrawee booboowhoo wirrawee booboowhoo
> yaaana nyiindi yaaana janawee
> Aboriginal women strong
> Aboriginal women strong
> Walk with you walk with me
> Janawee.
> Wirrawee booboowhoo wirrawee booboowhoo
> Yaaana nyiindi yaaana nyiindi Janawee.
> (Walk with the sound track, Jacinta Tobin, performed by Jacinta and Leanne Tobin, www.jacintatobin.com.au)

Leanne and Jacinta introduce themselves as descended from the land of the Boorooberongal and Wumali clans of the Darug peoples. They describe how their Aboriginal heritage was lost to them until they were in their thirties, and since then they have made a lifetime commitment to learn and share everything they were denied of their Darug heritage. They share a strong connection to Darug Country in their cultural performances, not an imaginary pure land of a distant past, but the urbanised riverlands, the country permeated by the rivers and their tributaries that are the veins of Western Sydney today. The walk from Emu Green to Yellowmundee represents their connection to this river Country, which they express in the Song Sings On.

The Song Sings On
They've cleared all the trees
All but a few, once a woodland and plains,
Now suburbs and drains
The creek hidden from view
But the song sings on
Beneath the culverts and the streets
The song sings on
Echoes from past gone
Connected and proud
The Darug breathed one with the land and its song
Its mother, its child
But still the song sings on
Waianamatta Mother Creek
The song sings on.
The Cumberland Plains hides secrets untold
Sit listen and learn
To the call of the Earth
Her wisdom so old!
So still the song sings on
Its needs denied for so long
The song sings on
Now threatened and tired,
The waters so slow
Sing a watery dirge, strangled, submerged
The life gone from its flow
So the song sings on
Hear its voice beneath your feet
The song sings on.
Now is the time, the woodland can't wait
To right the old wrong
And strengthen the song
Before it's too late.
The song sings on
Beneath culverts and the streets
The song sings on
And still the song sings on
Wainamatta, Mother Creek
The song sings on.

(Walk with the soundtrack, Leanne Tobin, performed by Leanne and Jacinta Tobin, www.jacintatobin.com.au)

Walking with the Song Sings On is to walk to the sound of the Earth's song, a song of resilience in the face of overwhelming adversity, the song that sings on, calling the human to be connected to Country, so that once again the land and everything in it can be sung into well(-being). It is this song that I take with me as I leave Emu Green for the long hard walk to Yellomundee National Park.

Walking the songlines of the singing painting river

It is an early Sunday morning when we set out early to walk the Nepean Trail, walking along the riverlands from my home at Emu Green, towards Yellomundee Reserve. After mapping the 9-kilometre, up and down walk we decided we would only make it to a half-way point and back, not knowing where that would be but knowing we would know when we got there. On a second day, some weeks later, we walk from the other end, setting out from Yellomundee Regional Park, walking to the same halfway place and back to Yellomundee. Each time I walk with the echoes of the stories of this place, its songs and artworks, as I negotiate its rugged terrain.

In walking the singing painting river, my body moves through Country, learning to know its terrain, its air, water, minutiae of flora, insect and bird life. The walking is so arduous that words are dispossessed of their power and the body enters into a meditative state of images sounds, rhythms and movement. I write on my return from still photos, videos, and images that stay with me in my bodymind. The track is rugged with loose sand and stones, easily dislodging one's footing and the terrain is made of a deep gorges cutting into the ridge as water makes its way towards the river. Down down down we go, slipping and sliding on the rough stony ground then up up up we walk, struggling to catch our breath to get enough air to reach the top of the next incline. All the way I am concentrating on the ground beneath my feet, avoiding loose stones that will throw me off balance, slippy sliding down the rugged slope. I become deeply immersed in the colours of the ground, in the shapes and patterns of rocks, and pebbles, tiny changes of form that enter into my being as I become with the songlines of the walk. Each time we walk to the same halfway point, a tall spreading tree, and turn around, following our different tracks back to where we have started from.

Walk I, Walking in sandstone and ochre, 22 May 2017

Walking in ochre is meditation, movement, and terrain, rough stony ground formed from decomposing sandstone, massive ups and downs, high up onto stony ridge with vast rocky outcrops of apple green lichen lace patterned granite, hard and unyielding, then winding down into deep hollows where waterways have formed gullies that tumble down the mountainside towards the river. River always alongside, sometimes beside, sometimes far below. Deep clay colours of dark red, white, and yellow ochres make patterns along the road and across protruding rocks, artworks crafted on the rock of ages. Coloured stones and pebbles of all shapes and sizes add their aesthetics to the emergent images.

The walk is measured in creeks, up and down, up and down to five creeks from the only one named on the Google Earth, the closest to home, *Fitzgerald Creek* (wherever do they get these names?) to the halfway point of the great river red gum. Cold air hovers in creek gullies, green moss on ever-shaded rocks, some flowing with still pools of water and in the gullies bell birds chime symphonically. Birds call overhead, a melancholy crow, laughing kookaburras, carolling magpie, chattering rosellas, rarely visible, it is their song that accompanies our walk. And on the ground, ants, everywhere, the sisters busy with whatever ants are busy with at home. I stop fascinated at one huge triangular-shaped nest that colonises the entire dirt road to take photo and video, ants crawl up my legs, colonising my body.

There are no words recorded on this long walk. It is the multiple patterns and colours of ochre that stay with me, producing a strong desire to create with all of these images that I carry with me as I walk. I download the photos and make large colour prints of the images to find the words to bring them into text, and find that the river has created its own artworks and songlines that takes me right back to the deep time of the creation of planet earth.

Image #1, Sandstone
Delicate wave patterns
of pale grey, brown, light, and dark gold,
rise up in folds on grainy sandstone,
its lithic bulk contrasted
with filaments of finest green grass.

About 4,600 million years ago in the Cryptozoic and Phanerozoic eras, igneous rocks were produced from the volcanic eruptions of the Earth's still molten core. From these very beginnings, natural forces of wind and rain wore down these rocks, carrying the sediments into the seas and back onto the land over and over again. The sediments became compacted and altered, forming the sedimentary rocks of the Earth's crust. In processes of endless change as land uplifted and then eroded, sedimentary rocks were being reduced to sediment again and new sedimentary rocks were formed. The terrain where I walk the singing painting river is produced by these continuous cycles of rock formation as the land was uplifted and then eroded.

Ripple marks on Hawkesbury Sandstone[*] map the creation story of the ancient rocks formed from the eruptions of the Earth's core, the erosion of the rocks over eons, the breaking down of the rocks into grains of sand washed into the oceans and returned to the surface of the planet. Over millions of years these sedimentary sands were consolidated into sandstone up to 50 metres thick to form the bedrock of Sydney's Cumberland Plain today. The Lapstone Monocline, the place of this walk, rose up from the plains to form the ridge that travels along the river and far to the north at Wollemi, a place of ancient trees, rock stone and ochre. The river's sandstone artwork sings this ancient song that continues to emerge from the very ground on which we walk today, pulling us back into the deep time of the Earth's creation.

Image #2, Ochres
Images of ochre patterns
are exquisite artworks from
the track beneath our walking feet.
The first deep orange-gold
mottled in delicate patterns
of deeper red and criss-crossed
with scatters of stones and rocks
of all colours shapes and sizes
erupting from deep orange surface
and casting their own small shadows.

An erupting fold
of multicoloured rock emerges
tumbling down the steep track
against a multicoloured substrate of sand, stones, fine leaves,
intricate lines and colours
yellow, mauve, brown, red, orange, white, and grey
scored across and through
with fine lines in a chaos
of interwoven patterns.

The creation story of ochre, the ochre that marked the patterns of unknown totemic relation in circular patterns on my skin belongs to the time of the formation of the planet, the endless erosion by wind and rain visible in the merging of rock, sand, and colour. Ochre is a mineral, a clay earth pigment, a mixture of ferric oxide and varying amounts of clay and sand, derived from the ancient clays that compacted to form the bed rock of sandstone. Ochres range in colour from yellow to deep orange or brown and red. Red ochre, the most highly sought of all the colours, contains larger amounts of hematite or dehydrated iron oxide. Minerals in Australian rocks show them to be some of the oldest rocks on planet Earth with minerals derived from sediments 4,100 to 4,200 million years ago, formed into rocks that are more than 3,600 million years old (White, 1998).

Ochre is the oldest art material on the planet. Ochre is integral to Aboriginal Dreamtime legends, stories of creation and the traditional laws of Aboriginal people throughout Australia. It is the basis of ceremonial designs on the ground and on bodies. It sets down the Dreaming (*Gunyalungalung* in Darug language), the Creation time for the Aboriginal People which provides their identity and link to the land. Symbols painted in ochre on rock offer the first evidence of Aboriginal ethos, or philosophy, dating back more than 20,000 years, a thousand generations of people living in this land.

In contemporary Aboriginal art, ochre continues to resonate with this ancient past, incorporating song, music, dance, body painting and performance, intersecting in a particular place through its ceremonial reference points. The vibrant dark red ochre as the ground of Leanne's painting of Maria Lock draws on these ancient traditions, recalling the singing and dancing that once echoed throughout the river valley in ceremony. Singing and dancing the songlines in ceremony are the always unfinished business of singing the Country, recreating the timeless death of the old, and the coming of the new, in the cyclical returns of life.

Tired out now we have gone as far as we can and make our way to a massive spreading river red gum that lies ahead of us. The river red gum is the spirit tree of the rivers, the spirit trees of Leanne's painting. It is a place to camp, to eat our picnic lunch, to rest, and recover. It is *ngurra*, one's Country, home, stopping place, the same word in Pintubi, Luritja, Warlpiri and Darug.

> ***Image #3, Ngurra: river red gum place of home***
> Tall tall tree reaches upwards to sky
> ragged branches radiating out like arms
> decked with fine eucalypt leaves
> artwork is in trunk's
> display of gorgeous colours
> twisting diamond patterns
> pale greeny grey, dark brown, light pinky gold
> reaching all the way up to
> a blue blue sky from
> its base at the ground, a mound
> of discarded sticks and leaves
> a perfect place to camp
> a stop resting place created by
> Eucalyptus cameldulensis
> river red gum of river spirits
> a creation storyline as old as Gondwana itself.

In the beginning, the land masses of the world were aggregated into a single supercontinent called Pangea which eventually separated into two parts: Laurasia to the north and Gondwana to the south. Australia was part of the great Gondwanaland connected to Antarctica, South America, Africa, Madagascar, India, and New Zealand (ibid.). About 50 million years ago, Australia became the island continent that it is today. Plants, however, remained from its earlier links with Gondwanaland, the characteristically Australian vegetation derived from the ancestral Gondwanan flora. From trees of all sizes and shrubby mallee, occupying habitat from the snowline of the mountains to the shoreline with roots in salt water, in deserts and in swamps and floodplains, eucalypts have come to symbolise Australia (ibid.).

Eucalypts evolved with fire and have influenced the fire ecology of the Australian landscape more than any other plant group. Before the arrival of Aboriginal people (60,000 to possibly 140,000 years ago), the fires which moulded the vegetation were caused by lightning and other natural phenomena. With the coming of hunter-gatherers the incidence of fires increased as they developed their own fire regimes, evolving alongside the fire-prone flora already in place. The eucalypt's ongoing creation storyline is intimately entangled with the original humans in this land. It is the eucalyptus tree, our *ngurra*, that links the two walks, a place of arrival and rest. After resting, we turn around and follow the same pathway back to Emu Green to dwell in the iPhone images, still photos, and videos. It is only then I can begin to make a storyline of the walk.

Walk II, Walking in air, light, water and strange red fruit, 25 June 2017

We leave Emu Green and drive across the Nepean River to the other side, following the line of the river until we cross over the Hawkesbury River and drive up the windy road of the Lapstone monocline until we reach Yellomundee Regional Park. The park continues the traditions of its namesake, managed by Darug people who practise and teach fire regimes, the ancient method of harvesting the land that has evolved alongside the eucalyptus. Setting out from Yellomundee once again we walk along the ochre-coloured track but the shapes of the land are different and soon we come to the waterways that wind and criss-cross down gentle slopes and cascade beside us as we walk. Their small valleys are packed with plants of all sorts, including the ancient Macrozamia with its full flush of bright red fruits that is only seen in scatters along the path on the other side. This walk joins the two sides, puts the whole picture together, walking from and with Yellomundee.

This easier gentler walk is written on my body differently in the sense of body's expansion into air light space colour, deep blue of sky, serendipity of shadows on little creek crossing, water running through shadow of my body. Body longing to immerse in this clear clean cold water but we need to keep on walking. Gradually the terrain changes into the ups and downs of deep gullies forged by water as we approach the halfway point, the giant forest red gum, the place of our camp site, where the two halves of our walk meet up. We sit and picnic there to regain our strength then turn around, walking, walking, and walking, body returns again, remove last layer to light cotton singlet, feel air on skin, warmth of sun, then cool cool in shade as damp cold air of creeks caresses skin. Heading back towards Yellomundee, the embodied walking of the Yarramundi songline, a walking trail of sound, music, colour, pattern, in the singing painting river's creations. I want to paint this walk by soaking the bark of the Ironbark Eucalyptus but unable to remove the tough coarse bark, I create with playful images instead:

Bodies of water #1
Stream cascading down the clefts
in mountainside
tumbling over chaos of rocks
big rocks covered in bright green moss

small rocks tumbled into big ones
rocks visible under clear pools
bubbles of white froth foaming up
as water rushes through rocks
towards its river
shaping its way through ancient sandstone.

Bodies of water #2
Water making art with body
as body's shadow is captured
within its cold flow
flowing stream creates transcorporeal
bodies of water
shadow image of body unexpectedly appearing
within cascading stream
water flows through and of this body
that is seventy per cent water
head decked with bright green ferns
leaning towards water
trails of white bubbles flow in and through belly
at feet tiny green vestiges of plants stranded
on water-strewn pathway.

Bodies of water #3
Water artwork of deep still pool
reflected images of blue sky
and faint tiny white fluff clouds
visible through tall trees
trunks, leaves and branches reflected against
bedrock of pale green sandstone
pool has captured blue of sky Country
and puff of white clouds and tall trees that reach up to sky
a creation of upside-down world of Antipodes
the great storyline of multi-dimensional Country.

The alchemy of water is a creation storyline that began in deep time when the Lapstone monocline rose up from the otherwise flat Cumberland Plain of the Sydney Basin and water carved its way through rock. Water, always running down, across landscapes, through the eons and the ages, carving creases and wrinkles on the face of ancient continents. Lifegiving, the essential resource for survival on a living planet. Over millions of years streams eroded deep gorges and cliff faces back into the uplifted block on the coastal side of the divide. Their creation story tells of the relationship between the lithic materiality of rock and stone, and the powerful gentle fluidity of ever moving water that flows from its source high in the ridges on its travels to the ever-present river below creating its own microclimates. The ancient macrozamia plant flourishing amidst its moisture laden ecologies dotted all along this watery half of the walk.

Macrozamia #1
Scattered on grainy grey sand
of walkway
surprise of brilliant red fruit and
strange aqua green triangular outer shell with
long lime green spike
abandoned among
discarded pieces of stones and rocks
all shapes colours and sizes
light grey, dark grey, palest of yellow ochre
deep purple umber.
Stone and plant becoming together

Macrozamia #2
Surprise of bright red
draws gaze into splayed glory of
green fractals in perfect formation
of palm fronds splayed out from
central core with spiky leaves reflecting
criss-cross of light and shade
protecting deep within
its clusters of scarlet of red fruit waiting
to be dispersed
across the watery landscape
where it can grow its new colony
of ancient cycads.

The macrozamia (*Burrawang* in Darug) is a cycad that evolved during the Jurassic period (from 208 to 144 million years ago) when the Earth was uniformly hot and wet worldwide.

Australian cycads coexisted with megafauna marsupials, such as Diprotodon, and herbivorous flightless birds, such as Genyornis. Because of their gigantic size, megafauna were able to eat the toxic fruits of the macrozamia and provide a means by which they were dispersed across the landscape where they could establish new colonies. Eventually the megafauna became extinct, thought to be influenced by increasing fire regimes after the arrival of Aboriginal people. With megafaunal means of dispersal gone, new storylines are emerging about a link between the brushtail possum and the toxic red fruits of the macrozamia. Because of their small size the brushtail possum feeds only on the non-toxic green outer hull, leaving the scattered remnants of bright red fruit behind as a new storyline of ongoing recreation in the riverlands.

Walking and walking we climb higher up the ridge, walking towards the sky, the landscape turns more dry and stony as up up up we go, tired now until finally we catch sight of the tall spreading river red gum in the distance, *ngurra*, our stopping place. We have nearly reached halfway.

The river is always there

The ever-present river: fragment #1
The river is absent on this long walk until
relaxing at the forest red gum home place
against the broad base of trunk
on mound of discarded twigs and leaves
the tree that represents two parts of walk
brings the river into presence
visible as patches of clear bright blue
a long way below
in between thick stands of eucalypts
the river that is always already there
long before us and long after we are gone
created during the early cretaceous period
of the Sydney basin the river that
forms the veins of Western Sydney today
its many tributaries criss-crossing its urban landscapes.

The ever-present river: fragment #2
The Nepean (Yandhai) becomes the Hawkesbury (Derrubin)
at confluence of Grose River at Yarramundi
spans traditional lands of several Aboriginal language groups, the Dharawal, the Darug, the Gundungurra and the Kurindji.
Aboriginal people used the river regularly
fish traps could be seen at Yarramundi before sand and gravel mining
Gundungurra resisted the taking of their lands
continually applied for official ownership
claims failed.

The ever-present river: fragment #3
Thirlmere Lakes area was once part of the Nepean River
but because of the evil actions of the rain maker
it was actually taken away from the Nepean River
and the Nepean River was taken away.
In those days the Nepean River, known as the Wirra-Wirra
ran into the Parramatta River
then there was a huge flood because of the rain maker
and when the flood receded

> the river no longer flowed
> into the Parramatta River, the river of eels,
> it flowed into the Wollondilly
> but the Thirlmere Lakes area was cut off completely
> because it was an important [ceremonial] increase area
> very important never supposed to be touched
> it was for medicine plants
> and the legends tell the Thirlmere Lakes area
> is the place for new plants to come from.
>
> (Dharawal elder, Francis Bodkin)

The Nepean River today is the basis of an urban ecology passing through increasingly dense sites of urban agglomeration. Fish and other aquatic creatures, birds, reptiles, insects, mammals, plants, all living things can only thrive in these urban ecologies because of the corridors of connection created by the waterways. Every day on the Nepean River is a new day: swans have come and gone with the passage of winter into spring, red-browed finches twitter all round, a bowerbird exhibits its ever changing collection of blue objects to attract his harem, tiny silver fish disturb the still surface of the river as they jump for insects, eastern water dragon lizards become territorial, a dark swamp wallaby passes by, wattles come into bloom, in an endless proliferation of life in this urban commons. Now more than ever new songlines, storylines, ways of being and knowing are needed to sing the places of our waterways into well-being. The song sings on as the Hawkesbury Nepean River asks its question, how can I bring the river to you as the reader?

References

AILA (Australian Institute of Landscape Architects) (2017). The 3rd City: International Festival of Landscape Architecture. Sydney, Australia, 12–15 October.

NSW (National Parks and Wildlife Service) (2009). Yellomundee regional park plan of management. Melbourne: Department of Environment and Climate Change NSW. Available at: www.environment.nsw.gov.au/-/media/OEH/Corporate-Site/Documents/Parks-reserves-and-protected-areas/Parks-plans-of-management/yellomundee-regional-park-plan-of-management-090510.pdf (accessed 20 January 2019).

Rose, D.B. (1996). *Nourishing Terrains: Australian Aboriginal Views of Landscape and Wilderness*. Canberra, Australia: Australian Heritage Commission.

Somerville, M. (2013). *Water in a Dry Land: Place-Learning Through Art and Story*. London: Routledge.

Somerville, M. and Power, K. (2016). Meeting of the Waters. With Badger Bates (Paakantji), Treahna Hamm (Yorta Yorta), Daphne Wallace (Gamaroi), Leanne Tobin (Darug) and Chris Tobin (Darug). Summer Institute of the Antipodes, Western Sydney University, Australia.

Tait, N. (n.d.). Available at: http://darugweavers.tripod.com/ourblackandwhitefamily/id2.html

Tobin, J. (2017a). Proceedings. The 3rd City. International Festival of Landscape Architecture. Australian Institute of Landscape Architects (AILA),Sydney, Australia,12–15 October.

Tobin, L. (2017b). Old yet untold: hidden stories of this place. Exhibition catalogue. Springwood, NSW, Australia:Braemar Gallery.

White, M.E. (1998). *The Greening of Gondwana: The 400 Million Years Story of Australia's Plants*, 3rd edn. East Roseville, NSW: Kangaroo Press.

Further reading

A number of other sources were drawn on to create the words of the deep time storylines. They are located at the links included below which were all accessed in January 2019.Alaimo, S. (2017). Your shell on acid: material immersion, Anthropocene dissolves. In R. Grusin (ed.), *Anthropocene Feminism*. Minneapolis, MN: University of Minnesota Press. pp. 89–121.

Barad, K. (2003). Posthumanist performativity: toward an understanding of how matter comes to matter. *Signs: Journal of Women in Culture and Society*, 28, 801–831.

Hill, R.S., Beer, Y.K., Hill, K.E., Maciunas, H., Tarran, M.A. and Wainman, C.C. (2016). Evolution of the eucalypts: an interpretation from the macrofossil record. *Australian Journal of Botany* 64(8), 600–608. Available at: www.publish.csiro.au/bt/BT16117

Karskens, G. (2010). *The Colony: A History of Early Sydney*. Crows Nest, NSW: Allen & Unwin.

Macphail, M. and Thornhill, A.H. (2016). How old are the eucalypts? A review of the microfossil and phylogenetic evidence. *Australian Journal of Botany*, 64, 579–599. http://dx.doi.org/10.1071/BT16124.

Somerville, M., Tobin, L. and Tobin, J. (2017). Connection to Country. Plenary presentation, The 3rd City: International Festival of Landscape Architecture.Australian Institute of Landscape Architects. Sydney, Australia, 12–15 October.

Springgay, S. and Truman, S. (2018). *Walking Methodologies in a More-Than-Human World*. New York: Routledge.

Sundburg, J. (2014). Decolonizing posthuman geographies. *Cultural Geographies, 21*(1), 33–47.

Terry, I., Walter, G.H, Donaldson, J.S., Snow, E., Forster, P. and Machin, P.J. (2005). Pollination of Australian macrozamia cycads (*zamiaceae*): effectiveness and behavior of specialist vectors in a dependent mutualism. *American Journal of Botany, 92*(6), 931–940.

5 Riverlands' watery ways

Introduction

Riverlands' watery ways ask the Hawkesbury Nepean River's own question, how to bring the river to you, its reader. It is a methodological becomings within the intimacy of the Hawkesbury Nepean River's watery ways at Emu Green. It explores how the river creates its places on the songline where water and land meet and everything is in a moment-by-moment relation with water, rock and stone, soil, weather, air and light, seasons, bird song, cicada call, plants, trees, litter, in its ever-changing becomings. A methodology of 'spots' is developed to engage with emergent representations of the sensory corporeal experience of the riverland's watery ways. The proliferation of still photos, small videos, and skeletal river notes are filtered through seven spots: *Bedrock; New swim hole; Regeneration; River's crossings; Litter; Drainway*; and *Spot of Death.* It considers the sensory corporeal experience of the riverlands where the human body is never one body but is also always in relationship with air and light, bird song, cicada call, rock and stone, soil, plants and trees, litter, all of the inhuman, more-than-human and humans that come into being moment by moment in relation with the riverlands' places.

> ***Riverlands, 10 May 2014***
> How can I capture the riverlands for you
> smell of wild fennel in sun after rain
> tinkle of bell birds
> rise and fall of cicada call
> clip clop of gumboots through
> wet of last night's dew
> walking to river's edge's familiar dank air
> of river wetness
> settle against large sandstone rock
> on casuarina needles overlooking
> deep part of river as
> light breeze riffles olive green surface
> and water catches sparkles of sunlight.

Moments of encounter

Every moment in the riverlands is different. There are seasons, weather, time of day, sounds, smells, movements, skin touch, vibrations, light, air, water flow, and the ways that all of the river's creatures respond to its ever-changing ecological sensorium. River texts slip out just as they emerge, images in the process of formation. Turning the proliferation of moments into a chapter that represents the whole as an emergent process seems an impossible task of representing the unrepresentable. Writing and language falter in the face of the riverlands' proliferations. I wonder if the senses could be a way to sort them, but every time I consider another river text fragment, there's something different. A passing glance at '*swallows flicker across still surface / body desires immersion*' proposes *Immersion* as yet another category. It is like trying to hold flowing water still, moving against the inevitable flow of the river's desirelines, until I recognise that each river text only comes into being from a single stopping place in the flow of the walk.

Points of emergence in river's methodological becomings

At each of these stopping places, the words of the river text come into being at the same time as I create still images and small movies as a multimodal record of the riverlands' becomings on my iPhone. The way I come to know and name these places is as spots, some of them already with their own names. When I come to think of the river's texts in relation to *Spots*, I can enter into each fragment of text and know where it comes from, its very material place of belonging. I print, cut, and sort, hundreds of pieces of text into folders according to *Spots*. Seven spots emerge and I give each one a name: *Bedrock; New swim hole; Regeneration spot; River crossings; Litter spot; Drainway spot*; and *Spot of Death*.

Spots now sorted, I enter deeper into the river's texts and contemplate the countless moments where everything is intertwined wondering how to write from the river's texts, how the spots are related to each other, and what the spots mean in relation to the riverlands' story of its methodological becomings. I walk to the river in search of answers, striding out of the house on my ordinary everyday walk. As my body eases into the rhythm of walking, feeling the ground beneath my feet, breathing deeply, letting go of tension, all of the spots are within me, coming and going in that body sense before language. I am walking the spots, walking the chapters, walking through the sequences, taking my book for a walk. Or it is the riverlands' book taking me for a walk?

Walking the book's songlines

Riverlands, 21 April 2019, 9.09 a.m.
Walk along drainway
spot of flowers
to spot of plantings
then river crossing
and new swim hole

journeys between spots of creation
where new stories spring forth
becoming a songline and spots are only
where all stories live simultaneously
in this becoming of the world
at the turning of the world's circle.

Songlines of becoming

As I walk, the river's spots emerge as places on a songline, places of creation of new/old stories coming into being over and over again. The spots are linked by the walking trails that the ancestors made as they walked through the landscape doing all the things of their everyday lives. Each spot is a stopping place, a special places of song and ceremony, a place where the world comes into being over and over again. My mind turns to thinking of the whole as I swim in my new swim hole, swimming my book this time, or my book swimming me, as it often does, water flowing over my body, taking things away, but this time I start to feel the loss, the loss of the whole, the loss of many of the river's texts over those years, so I walk onto Bedrock, the next spot on the linking trail.

Becoming with the spectral

Just near Bedrock there's a sunny place at the base of a casuarina right against the riverbank where I love to sit, and as I sit relaxing in the warm sun, the spectral figure appears from her usually hidden place, the grey figure of a woman, head bent over, dressed in familiar long dress and bonnet, a spectral figure of colonial days. Sitting in this warm calm place I wonder if I might find lost river texts if I use *spectral* as the search word? At home, I type in *spectral*, and many lost river texts appear. She also gives me photo logs of images, taking me to the real multimodal nature of this river writing.

Multimodal images

The gift of the photo logs means I can print out the list of photos and videos with their dates and content titles, and annotate them with the name of a spot. The spots offer a way into making sense of this multimodal data and another category emerges, as a space of *Excess*, the in-between spaces, the liminal, and immanent, that falls outside of the spots. The photologs lead me back to the photos and videos to contemplate their multimodal becomings.

Dream analysis

Night-time is waiting time, a time to wait for something to emerge from the chaotic space of unknowing after a night's sleep. At dawn, in the in-between space of waking and sleeping, a dream analysis emerges in swirling images from the photos and videos of the places of the riverlands. Each spot comes to life in its difference, of colour, movement, sound, affect, and speciation (how is the human made anew in this place?), creating an emergent world-becoming of the riverlands' places. In the following days images and words come to the surface in all the liminal spaces of half-sleeping, half-walking, swimming and walking until each of the spots on the songline creates its own song.

Bedrock

Bedrock's song is the tinkle of bell birds, twitter of tiny finches, cry of water birds, crystal-clear call and response of whip birds, and the song of the casuarina swishing in the wind. Casuarinas line the riverbank, creating the shape of this place of water, rock, stone and sky, a place of soul song and re-creation.

Bedrock's river text sensory ecologies

10 May 2018, 8 a.m.
Rising mist looms
across water's still surface
sun remains
hidden behind cloud
crow's lonely caw echoes as
white egret flies along river highway
with deep guttural call.
Body longs for immersion
but sitting against old crinkled trunk
cold air eddies around left shoulder flowing
from upstream in this early
cold morning of the world's beginning.
I'd much rather be in there than writing.
What does it mean to write the river?

Within *Bedrock* there are countless moments of encounter in the long slow winter of healing. At this particular moment, the changing season is becoming colder. Walking through frosty white grasses, the water gradually becomes too cold to swim, the air even colder, and mists rise from the river's surface as the freezing air meets warmer water. In these times I sit beside the water at Bedrock at the spot where, in warmer times, I would slide in, always desiring immersion. Instead, recording passing moments of the quiet river's flow, the surface of water, the river's skin. Calming ripples appear on the surface, bubbles and bits and pieces float by, balloon vine pods, and flowers, fragments of reed, and so on, a pleasurable randomness that keeps me enthralled with the river's silent flow. I notice the play of light and shade as the sun comes and goes behind the clouds. Then there are always birds, often announcing their presence only by sound, the whip birds call, bell birds tinkle, twitter of tiny invisible birds, visceral call of egrets as they fly overhead or swans honking as they come and go in their annual migrations. At other times birds insert themselves into my thoughts as they appear like the cheeky wren doing their bird things, and my thoughts flitter flutter and become bird-like. These are the only times I take my computer and write at the river, and I question the reason for writing, as well as the how, of bringing the riverlands into linear representation.

Bedrock's multimodal becomings

Every image of Bedrock is framed by the casuarinas, their leaning branches and fine needle leaves providing a stage for Bedrock's becoming with light, sound, air, and the river's slow flow. Bedrock's images are all about the play of water and light, the changing colours of the seasons, weather, and times of day. A pink mist rises and falls over a river of reflections at dawn, a black river in the dark of evening shines with silver ripples, a green river is quiet and calm under a cloudy sky, the river is bright blue on a clear sunny day, and silty brown when the big rains are flowing down. Silver sparkles on blue water glisten the colour of blue diamonds that sing in tune with the whip birds' crystal-clear call. Bedrock's overriding colour is gold, as sunlight and water come into play on the river's rocky bottom. Golden light glows from large flat rocks and river stones in the play of sun, water and rock. Ripples of gold and silver intersect in patterns of wavy lines in a dance of sunlight, water, rock and stone.

Bedrock's waters move slowly, an arena for floating bits and pieces of balloon vine, salvinia, leaves, flowers, that float slowly by, only disturbed in their passage by wibbles on the surface that emerge from some unknown perturbation below. Sometimes the river's flow makes its own sounds, a mild rushing when the brown flood waters flow by, a tinkling of rain on the surface of green water, but Bedrock's sound is bird song, the crystal-clear call of the whip bird, the high-pitched call of water birds, always the twittering of tiny invisible birds in the trees overhead. Its presences are birds too, there are always birds at Bedrock, ducks, egrets, swans, pelicans, tiny grebes, moor hens, cormorants, and the birds of the air, swallows that dart and flicker across the water's surface, tiny birds that hop from river stones to river stone dipping beaks into the shallow waters in between, their rocky supermarket. One photo shows a congregation of birds, three pelicans sitting on rocks mid-river near three cormorants perched on protruding sticks, two tiny birds hopping on river stones near Bedrock's tiny island, as a grebe floats slowly by.

Bedrock's affect is quiet contemplation in the elemental space between water, rock and stone, and the blue air of sky, the space between the materialities of the river and the vast unknown of the blue sky. A sinking into the miracle of its golden light, the dance of sunlight, water and rock. It is a place where the swishing of the wind in the casuarinas reminds one over and over again of the presences of ancient Aboriginal stories in this place, as they call the names of all who ever dwelled here since the beginning of time. It is the place of the spectral figure that haunts the riverlands with her reminder of the riverlands' absent presences. It is a borderland, a space between the great unknown and the planet Earth, that calls forth the work that the soul performs, the deep emotional work that takes place under the surface in the borderlands of life (Anzaldúa, 1987).

Litter spot

Litter's song is the song of camping, making a home place where the river's bank creates a sheltered enclave at the back and at the front an opening to a wide still pool of the river. There a circle of river stones marks the place of the hearth, it is *Ngurra*, Country, camping place, home place. It is a place for human dwelling with all of the artefacts that signify human presences in this very local place, and all over the world.

> ***11 January 2018, 6.52 a.m.***
> Empty red chair
> question mark on back
> sits facing
> the river
> in conversation
> at peace
> alone
> amongst the global materiality
> of litter.

The question mark on the nomadic chair symbolises the profound paradoxes of this spot which is both here and everywhere at the same time, arousing powerful affective responses in its ever-changing manifestations of the global created by the river's litter. First, it was the 'place of stones', small girls throwing stones into river water, splash and ripple splash and ripple, small soft hands smoothing out sandy dirt for a circle pattern of stones, strident child voices claiming the river stone fire as their own. Soon after that it becomes the 'fishing spot' as fishers leave behind corn tins, bread bags, standing lines held in forked sticks. Then it is a camp site, inhabited by many comings and goings, drink bottles and cans proliferate. Plastic chairs appear in a row facing the wide open waterway, a colorbond (steel fence) structure stands, decorated with colourful graffiti, and an equally colourful abundance of litter decorates the site. Soon litter owns the site and it becomes *the litter spot*.

The one thing that remains constant among all this change is the circle of large river stones, holding the charred remains of a fire, the hearth that signals home, it is *Ngurra*, a home place, a place of belonging. The front of the site faces a wide expanse of river, flowing towards the sea, a place for fishing, drinking, eating, partying, relaxing by a fire on a cold day at the river. The secluded enclave at the back of the site is hidden from view, a bushy path winds through to a wall embedded with river stones, the bank of the river in the time of the river's full flow before the dam. This hidden place is a perfect place to make a camp, a place of belonging, a bush shelter, a home place.

But it is litter that owns the site. Litter that comes and goes, creating its own emergent installations. Litter that has its own text that tells its own stories: *Nelson County Bourbon & Cola; Woodstock; Kentucky blended bourbon & Cola, original blend, 4.8%; Jack Daniel's, Old No 7 brand, quality Tennessee whisky & Cola, whisky distilled by Jack Daniel distillery, Lynchburg, Tennessee, USA*; Some of litter's text has its own images too, an empty cigarette pack lying on the ground shows an image of a damaged heart muscle accompanied by the text: *Smoking causes heart disease by damaging the blood vessels and blood supply to your heart.* Another empty pack displays an image of a premature baby, wrinkled, red skin, respirator taped on nose and forehead, disembodied hand cradling head, wrinkled eyes closed, and text that says, *Smoking during pregnancy reduces blood flow in the placenta and limits the oxygen and nutrients that reach the growing baby.* Litter's stories are of everywhere and at the same time powerfully local within human bodies, from the beginning of human life.

Litter spot's multimodal becomings

Images of the litter spot are strangely colourful and material, produced by the litter itself, scattered amongst the bushes, hidden in the enclave at the back, stacked in bundles in the stone fireplaces, or appearing on the surface of the river. A pile of cans and plastic bottles in colours of bright reds, blues, black, greens, yellows, white and silver, contrasts with the soft mottled earthy colours and textures of the river stones. One large round river stone stands out in its pale yellow ochre colour with a red ochre patch. Seven others are mottled deep umber on ochre, in sizes and shapes from large and small, round and oblong, all neatly placed to complete the fire circle that holds the massed glitter of litter.

The affect of the litter site swings between disgust, repulsion, anger and the attachment that led me to create a series of installations with the river's litter, rescuing litter from the river, and its places scattered among the bushes and in the hidden enclave at the back, piling it all up into a colourful installation that had its own migration tendencies, as each day when I returned, stuff had moved and changed. I moved some back again until finally, soul weeping for the rivers of the world, I requested the local council to bring in a truck and remove the litter. When finally this happened, I was devastated and surprised at the loss of the river's litter, feeling it belonged to the river, not some other place where it did not choose to go, but then in a matter of days the litter returned in its endless proliferation at the litter spot.

New swim hole

The song of the new swim hole is a song of stillness and movement, it is a song of longing, the longing to belong of the forever migrant, always on the move, imprinted with the memory of migration, and yet recreated every day in the stillness of the quiet pool. A still quiet pool lingers inside its wall of balloon vine, protected from the nearby rapid that bubbles and flows with its forever movement towards the great oceans of the world. A place where plants, animals and humans find stillness and rest before being on the move yet again.

Swim hole, 9 November 2018, 8.30 a.m.
Every day the same every day
different
blue dragonflies hover
cicadas loud in their drumming
no red-beaked moor hens today only
proliferation of tiny white flowers floating upside down
on patches of foam
tiniest pale green and white petals
cradling golden stamen within
ready to be fertilised to produce
ever present seeds of the toxic balloon vine
that creates new swim hole.

It's drought in this the driest inhabited continent on Earth. River levels are down and Bedrock is murky, furry scum rising up as body cruises up and down. Swimming ceases until one day my son leads me, secateurs in hand, to a spot long since lost to the invasive balloon vine. But it's winter, and drought, and the balloon vine hangs brown from the trees, opening up the faintest of trails to a hidden spit of land that protrudes into the river. A single large casuarina leans over a deep quiet pool that lies inside a cascade of rapid bubbling waters flowing across river stones in the middle of the river. The bubbling rapid cleanses, the water is deep and cool, the pool becomes the *New swim hole.* Body becomes aquatic, breathing in deeply at water's surface, breathing out forcefully into water, thoughts displaced in rhythmic watery meditation. River's bubbling sounds enter my body, as my body's breath becomes part of river's sounds. As I contemplate the vast mountains of balloon vine surrounding this private, hidden swim hole where I have never seen another human, it dawns on me that it is created by the noxious balloon vine.

As summer's warmth approaches, the small green fruits of the mulberry tree begin to turn from bright red to deep purple black. Day after day I harvest the river weed's gift, leaning out over the water or stretching up to pull down branches wherever shiny purple black fruit sparkles in early morning rays of sun. Colour bleeds onto hands, and I wonder how is colour produced, and what does it mean to eat colour? One day a grebe gathers mulberries with me, emerging from its hidden place in the riverbank, before dropping back down to the river and floating away. An eastern water dragon emerges from lower branches, eating mulberries, and we each move closer together until we can see each other's breathing and the flicker of an eye, becoming molecular, becoming imperceptible together.

A turtle is there one day when I arrive, its beady-eyed head extended on a long reptilian neck from its tough patchwork shell. It doesn't move at my presence, I wonder if it is alive or dead, so I reach out to touch it and it retracts its head as I gently lift it off the ground to reveal clawed reptile feet, and place it gently near the water's edge in case it is sick. The next day it is gone. A mystery snake skin lies out across the grass, shiny grey and dark brown, a brown snake. I feel its fine papery skin and watch it over days as it disintegrates by the swim hole where the snake has slipped into the water so it could shed its skin. We are all here from time to time but we keep to ourselves, the new swim hole is a solitary private place of being alone.

New swim hole's multimodal becomings

The colours of the new swim hole are the colours of movement. The ripples are always flowing gently from the bubbling rapids towards the still quiet pool. The sound is movement too, the bubble of rapids, always flowing away, all the way to the sea, past the waters of first contact, past the little island in the big bay, all the way to the waters of the Pacific Ocean that brought my old people to their new home. The dreaded weed *salvinia molesta* finds its refuge in this quiet pool too. In spring, it begins its process of vegetative reproduction, leaves break off from the parent plant, floating by, searching for a place to stop, a place to belong. Some arrive at the edge of the water hole where they settle and produce tiny clusters of white flowers. We are all migrants here searching for a place to belong.

The grassy landing place is a place of absent presences and camouflage. The lizard mimics the colours of the fine green grass that covers the tiny knoll, changing colour as it changes position, orange and light-coloured patches and stripes changing to the greens and yellows of the green and dried grass. The turtle is a dark brown rock, the colour of the bare earth where it stubbornly sits at the base of the tree, the snake leaves only its skin as a sign of its absent presence. None of us here at this private place want to be seen, all camouflaged by colour, and silent, in the sound of the bubbling waters that flow to the great oceans of the world.

Regeneration spot: trees, weeds and tender intimacies

Regeneration's song is the song of seasons, the seasons of the wattles coming into bloom, of plants dying down in the winter, of the times of the annual fish spawning. Its rhythm is the slow time of a tree growing, the timing of becoming with the intimacy of plants, time of caring, the rhythm of the slow time of the planet.

> ***7 May 2015, 11 a.m.***
> Planting
> knees on twiggy ground
> bark with lichen worm holes
> fine needles sticks wood close up
> bits of plastic tin glass
> watched by white egret
> from distance in quiet

smoke gazed hazed river
ducks cruise, feeding on layers weed and scum
rivers skin, or only sign of
pollution, river dryness.
Talked about silver fish at dusk
with fisher Alan:
herring he says, jump out that far (with hands) *catching insects*
all at once jumping silver
saw one catch insect foot above water jump and snap
you can eat herring, they crush the bones
'Like sardines?'
Same family, caught them once for fun, only grow that big (hands again).

Planting, weeding, repairing, watering, eating, talking, the monthly Bass Fishers bush regeneration meeting is a time for tender intimacies between strange bedfellows. It is time for the planet. Slow time. The time of watering trees with a watering can from the river. Or the time of a tree growing. It is a time when the aquatic and land-based creatures meet and intersect, organising their humans into tender intimacies. A tiny spider hurries across the ground of a plastic tent that protects newly planted trees. Each tent a mini-eco-system, droplets of moisture condensed on their sides running back down to the parched soil at the base. A damaged acacia with a small bandage around the broken place and a thick straight stick placed as a prop to hold the tree upright while it repairs. Mosquito fish emptied out of watering can back to their watery element, and gudgeons at river's edge announce the coming of the Bass fishing season.

In return for access to the waterways for their catch and release bass fishing pleasures, the Club is allocated a riparian zone to repair, replant and care for. They have been tending the riparian zone at this spot for eight years. Seven years ago I joined them in the endless task of planting, weeding, watering, repairing damage, and community building through shared hard work, rewarded with relaxation over a traditional BBQ lunch. Sitting against a wide casuarina trunk on its mound of ancient pine needles, the conversation is equally about fishing and plants. The talk turns to the latest fish catch, the hot spots where fish crowd for the catch. Together we share our despair about the violence of four-wheel drivers who take pleasure in ramming our trees from time to time. Or about the impact of a long hard dry winter when I thought many of our trees had died.

They told me to keep watering them, but I wanted to pull them out. Then in spring the rains came and the dead twigs turned into new fresh leaves sprouting vigorously from the base or up and down the stem, depending on their habits. I talk to them, *I thought you were dead*, tending, cutting off their dead twigs to make way for new growth, but most of all, I learned their names, both common and Latin names that the true regenerators use: Cheese tree: *Glochidion ferdinandi*; Australian indigo: *Indigophera Australis*; Coffee bush; *breynia oblongifolia*; Forest red gum: *eucalyptus tereticornis*. Regeneration, coming back to life from death.

Another time a strange sight strikes me as I walk past our plantings, a massive casuarina branch has fallen, breaking one of the tall thin sapling eucalypts in its fall. Casuarina is still fully alive with its green grey needles and mass of brown prickly seed, all forming a playground for little finches, bright red under wings and around eyes, flickering in and out with light twittering sounds. I follow the line of the broken tree to its main trunk, wondering if this is another act of vandalism, but following the line of the trunk, I find it broken off because it is completely rotten inside, eaten by termites, or maybe grubs of some sort. There are large and small holes in the inside of the remains of the trunk, powdery remnants of decay, a whole other world, the crumbling debris of animal life. Life and death in the tree world intersect, bringing new worlds into being.

Regeneration's multimodal becomings

The all-pervasive colour of the regeneration site is green but there are so very many different shades of green. The grey green of tall eucalypt trees that reach up to the sky; the dark green of casuarina needles feathering above the lower bushes. The bright green of the small round leaves of indigophera and breynia, especially vivid when their new leaves return with the spring rains; the light green tinged with pinky red of new bottlebrush leaves, and more. Against this green, there is the surprise of colour. Sometimes it is the flowers, the brilliant fluffy yellow of the wattles, the gentlest palest pink of the native hibiscus, all designed to attract their pollinators. The fluorescent blue of one butterfly, the brilliant red, yellow and black of another, and the brightly coloured birds that flick and flitter through the bushes.

There are images of fungi and lichen too, a bright orange of fungi contrasted with the brown of a dying limb, a large multicoloured plate of live fungi circles of light grey, pinky brown, pale yellow and dark orange, bright against the more delicate shade of pale, grey green, lichen lace on a dead tree branch. The cycling and recycling of colour form the story of regeneration, the reproduction through pollination by brightly coloured birds, insects, butterflies, but also the brightly coloured fungi that turn dying plant material into nourishment for the next generation of plants. It is all about regeneration of all living things, regeneration of the human too, producing new humans in tender intimacies with the care of plants.

River crossings

The song of the river's crossings is the song of play and of crossings over to the other side. It is a song of improvisation, of the in-between, the frictional space where difference meets, a space of possibility and imagination, where the human can become other to themselves and experiment with singing the songs of other living creatures who inhabit the crossing, all of them coming into contact with each other, changing and becoming other to themselves.

Riverlands, 10 September 2013, 12.30 p.m.
Underneath this grass is my baby eggs
we are so smart, we're just like humans.
we're sand birds
birds that live in the sand.
make nests in the sand.
our eggs are hidden under the grass.
we don't sit on them
they hide under the grass
and we make fires right next to them
so they stay all warm.
We think like humans and here's our collection of herbs
and I need to put rocks around them
so they can't blow away
we really need so many rocks, good rocks
we made the property so no one could attack our eggs.
we don't have a boss, it's just the way we work
it's just the way all sand birds, it's just what they do
they make walls around their houses
we all share this area cos' we all like this spot.

The river's crossing spot is named for the four-wheel-drive vehicles that love to drive through the river at a spot where the water is low and a bed of river stones is close to the surface. Massive sandstone blocks are placed at the edge to prevent four-wheel-drive vehicles from driving across the river to the other side. Winches are used to shift the blocks, tracks are gouged through wet ground to make a way around them. There's many stories of what happens on 'the other side', dope smoking, dope growing, pig hunting, a cop was once shot at when he tried to investigate.

At the river's crossing, everything is possible, it can be singular or plural, coming to represent the multiple crossing-overs made possible in this place of transgressive play. Humans become animals, animals become human, humans become elemental with mud and water and river becomes infused with human play. It is a place of human habitation, a cross-place of invasion, a crossing place where anything is possible. A turtle is caught on a fishing line by mistake, the hook carefully extricated from its mouth and it runs as fast as turtle legs can carry it back to the shallow water. Gracie learns to count stones, arranged in a row, in Vietnamese and English, while her father fishes. Bella sees a mermaid, Ollie a crocodile, as they cling to my back and neck to play in the shallow water at the river's crossings.

A flooded river leaves layers of mud everywhere and Lulu and Charmaine make a mud kitchen, Lulu picks up mud with two hands and flings it towards water, chanting to the mud '*and then some humans, was trying to get the mud to go away, was throwing the mud away*' as she leans forward again, picks up more large handfuls of mud and throws with both hands sideways towards the river. Lulu says I am 'human' but the girls are as much mud and water as human, their actions, vocalisations, imaginations, and words, simultaneous with intra-action of mud and river at the very edge where water and land meet and merge.

The river's crossing is so powerful in its imaginative potential that it pulls into its own spot all of the multitude of crossings over, becoming the metaphorical site for all becomings other to oneself in its immediate surroundings. The river camp, seal island, the boulder site. At the boulder site we feed [the river] rocks, ride our spaceship, float little balloon boats, a mother and a baby, have a hair-dressing salon, a water picnic, and become rock jagwars [*sic*] with lairs of sticks. And, of course, on seal island, we become seals, our homes on large flat stones facing the smooth open space of the river where we sing songs and tell stories as we go to sleep on our seal rocks. The river's crossings is a place of endless becoming other to ourselves in every possible and impossible way.

The river crossings' multimodal becomings

The colour of the river's crossings is the colour of dirt tracks, sometimes dark and muddy, other times light and sandy. Ever-present tyre tracks inscribe the story of movement, coming and going, backward and forward, wheeling around until the tracks are lost in the stony waterway of the crossing where cars disappear around the little island to the other side. The shallow stony waterway makes its own distinctive sound, a slow gentle gurgle as the water flows over river stones, something like a resonance of your own circulatory system, a deep-pitched murmur that is already within you, It is the sound of one's own heart, echoing other memory stories of frictional encounters in the space between water and land. It is these frictional encounters that produce the multiple affects of the crossing site, from the fun of play and the exhilaration of difference, to visceral responses of anger in the productive discomfort of the contact zone.

There are images of multiple encounters between birds, insects, animals, and human in this crossing place. The turtle caught on the fishing line, then running back to the water. The ducks cruising slowly back and forth in search of bread scraps. The swamp wallaby visible only in fleeting glimpses as it disappears into the bushes when it sees the humans arrive. Ravens hover in the trees above, always searching for the remains of food and landing on the ground when there is something that catches their eye. Images of young children entirely comfortable and at ease in their inter-species and inter-elemental inhabitations, becoming bird, becoming seal babies, becoming animal, becoming mud, becoming with river stones, always singing the song of difference.

Drainway

The song of the drainway is:

> The Song Sings On
>
> They've cleared all the trees, Now suburbs and drains, Creek hidden from view, Beneath culverts and streets, Now threatened and tired, the

> waters so slow sing a watery dirge, strangled, submerged, The life gone from its flow, The song sings on.
>
> (Leanne Tobin, www.jacintatobin.com.au)

It is the song of Lapstone creek converted into a storm water drain through which flows all the detritus, debris, and litter of stormwater drains, liberated into the place of a winding slow waterway where flowers create a wild proliferation of colour, form, and beauty, singing a song of red, gold, purple, pink, white, and maroon, as the drainways song sings on in the colour of flowers.

> ***Drainway, 3 November 2017, 8.50 a.m.***
> Walking after rain passing multiple new
> installations
> chicken wire intertwined
> with long tree branches
> and criss-crossed all through
> with leaves twigs and plastic bags.
> Multicoloured tin cans drink bottles
> and brightly coloured plastic bags
> arranged in a tumbled pile.
> Discarded shopping trolleys becoming
> more embedded
> with drainways debris making
> new home for critters
> to live in this edge place.

Over and over again the drainway makes and remakes its own installations, using everything that turns up in the flow of the storm water drain that runs along Emu Green and flows into muddy pools of stagnant water that rushes through in torrents when it rains and dwindles to pools of grimy sludge in between. A paradise for rubbish, and weeds, especially the noxious coral tree, I wrote once to the Council suggesting that they could turn the drainway into a reed filtration system. It didn't happen. Instead the drainway created its own perfect litter installation. A beautiful, complex multi-textured, multicoloured intermingling of natural and artificial materials appears after heavy rains. All is held within intertwined entanglements of twigs, branches, leaves and logs arranged against the concrete causeway that spans the canal of the drainway. Colours are soft beige browns with flecks and patches of delicate faded red and orange where bark remains on peeling logs. Placed among these are the textures and colours of pale grey plastic bags, scattered shapes of worn white polystyrene, fragments of vivid orange plastic, faded blue of cardboard carton. It is as if everything is made more gentle by its slow becoming with water's flow.

It wasn't until wandering the drainway searching for a way to express grief at the massacre of 51 praying people in New Zealand that I even recognised the full beauty of the drainway's flowers. Wandering along the drainway from Emu Green, as I do every day, pondering on what flowers do I want to bring, the vivid red of the coral tree flower inserted itself into my consciousness. Close up so perfect, so bright, delicate and beautiful on a cloudy grey morning. And then the brilliant gold of another weed, the cassia bush, multiple golden heads gathering into themselves all the light from this dull day, and sending the joyous colour back out into the world. Further along, more flowers, a pathway of weeds draped in vines with large, soft, white-petalled flowers with a tiny splash of mauve on each petal, a floral arrangement all of their own. And then there was more and more, flowers I had never noticed, every colour of the rainbow, all dismissed as weeds. So the drainway became the spot of flowers. On the morning of the ceremony I walked early and gathered bunches of all the colours mixed together, intertwined with dark purple and pink morning glory, and translucent white daubed with mauve, flowers of healing, flowers from the rivers of eternal life.

Drainway's multimodal becomings

The drainway's colours are both the colour of its litter, garish and usually grotty with the dirt that flows with it, but sometimes gentle and delicate, washed by the water's flow. These colours are so different from the beauty of its flowers' bright red, gold, pale pink, purple, white with its splash of mauve and deep maroon leaves. The drainway is inhabited by all of the birds, gaggles of ducks dipping and splashing in its water, announcing their presence with loud, honking, contrasting with the myriad of peewees, magpies, and the shy white-faced blue heron. The drainway is multiplicitous, the most complex and changing of all the spots, a complex site of the most beautiful, multicoloured of flowers, and the most ugly piles of litter as the waterway makes its own litter installations. It is the quintessential urban, the confrontation of waterway and urban world. A place where the song sings on.

Spot of death

The song of the spot of death is the song of deep mourning and loss, the song of grief and loss for the planet and the song of joy in the gift of the riverlands of life. It is a song of life and death, where death produces life and life produces death in the eternal cycles of planet Earth.

> ***Riverlands, 30 July 2016, 5 p.m.***
> There are no words,
> only the image of beloved
> border cross collie
> lost and found again
> claimed by the river as its own
> bloated body lifted gently

to riverbank place beneath the rivers ground
under a protective covering of river stones.
The beginning and end of the river's book,
life and death in the Anthropocene.

The spot of death frames the beginning and ending of the riverlands book, its beginning in the time when there were no more dogs to walk in the riverlands and the river began to walk me to its spots, and its ending in the final chapter, Life and death in the Anthropocene.

The space of the in-between

And yet, as soon as all the spots are settled in their places, the multimodal data of this project generates another layer of dream analysis, dreaming in images of the in-between, spaces of the liminal, spaces of becoming. The grasslands as the quintessential in between, but there are many others too. Mushrooms, toadstools, and fungi growing in the liminal spaces between dead and alive. The muddy in-between of the flood lands and the water. The river's graffiti, incomprehensible but announcing their presence. Frosty grass glittering in the early morning sun, kangaroo grasses swaying in the breeze and the river's stones stranded on the flood plains. Stone stories. The architecture of balloon vines, changing the whole landscape. And sounds, the all-pervasive presence of bird song and water's gurgling and bubbling over rock. Living in the images of the night and in the morning dreaming time analysis. Swirling in that space in between images forming but not forming into words.

Grasslands

Feathery pink grass heads
lean over the pathway
kangaroo grass
sways in the breeze
love grass tickles softly
on walking hands
spider's web hangs silver
in morning dew and
tiny finches fly out
from grassy home.

The in-between has no colour, and many colours, in the swirling chaos of the world's becoming. If anything comes to symbolise the colour of the in-between, it is the colour of the grasslands. There's a triangular space of fine tall grasses with feathery seed heads, pale golden colour as soft as their touch as they brush my hands as I walk past. It is mostly *love* grass, a native of South Africa, the quintessential in-between, named from the Greek *eragrostis* from *eros* meaning love and *agrostis* meaning grass. In winter, the love grass sparkles white with frost in early morning light. A fine spider's web hangs silvery in between grasses lit silver in early morning dew. A flock of tiny red-eyed finches flies up to the branches of a dying wattle that sits in its midst. A cluster of bright green wattle seedlings spring up, the grasses are a nursery bed for the emergence of the new.

Stone stories

> Stone city
> stone supermarket
> languages of stones
> stony walkway
> stony riverbank
> grave stones
> shiny stones in wet of rain.

Stones are everywhere and every image of stones tells its stone story. Small girls make stone city on the riverbank, tall piles of flat stones, circles of stones, stones of every colour, size and shape. Birds inhabit a stony part of the river, streaked with white bird droppings as they peck and lift, peck and lift in between stones for their food of the day. Collecting stones for the garden creates a language of stones, gold ones, red ones, purple ones, blue ones, mottled multicoloured ones, smooth stones, flat stones, round stones, rough stones, big stones, small stones, from the stony pathway of the river's leftover stony river bottom before the dam was built.

Bird song

> A bird launches into its refrain
> all music is pervaded by bird song
> in a thousand different ways
> whip bird's call and response
> tinkle of bell birds
> twittering of tiny invisible birds
> squealing of black cockies as
> they fly overhead
> laughing kookaburras
> high-pitched call of water birds
> long lonely caw caw caw of crows
> echoes of profound loneliness.

Everywhere in the riverlands is pervaded by bird song all along the river and all the spaces in between. A lyre bird has never been heard here before but one day its insistent call announces its presence in the bushy in-between spaces of the riverlands. There is a cacophony of bird song. It's going to take a long time of sitting and recording over and over again so I search for the nomadic orange chair and place it carefully in a hidden bushy place close to the lyre bird's singing where I sit and record over and over again. Every time there is the same cacophony of birds, blackbirds, whip birds, bell birds, cockies, and the forever cheeping and twittering of tiny birds, all of them calling over and over again. From time after time of slow listening I learn to hear the sound of the lyre bird's mimicking of all of the others, just slightly off key, a tiny difference in song as it imitates the songs of all the birds in this in-between space of the riverlands.

Water's songs

> Soft tinkle of raindrops
> On still surface of water
> rush of water
> in river's fast flow
> roar of river's waters in flood
> splash of water over rocks
> bubble of river in rapids
> gurgle of river over river stones.

The river's watery song has the quality of the refrain, in its all-pervasive presence as the river splashes, trickles and roars in concert with bird call, cicada song, swish of wind in trees. The tinkle of rain dropping on the water, the rushing sound when the water is flowing fast, the roar of the river in flood, the play of water on rock and stone, sounds that enter my body as I swim, body's breath making its own gurgling with water.

River's weaving

Three images of the in-between space of the riverlands tell of the river's weaving. When the flood waters have gone, the river's weaving appears on the back of its casuarina trunks. Fine dried water weed winds in colours of pale gold and grey, delicate dried strands interwoven. Some of the weed is straight lines, curled and twisted around, other weed is tiny dried leaves, curly and crinkly, winding in and out, pale gold, orange brown, dark grey interwoven. The river's weaving spans the space from tree to tree, hanging in between in a woven thread of multiple river weeds. The river's weaving tells the story of the coming together of the becomings of all the different spots, linking the songs of *Bedrock, New swim hole, Regeneration, River's crossings, Litter, Drainway*, and *Spot of Death* in the liminal space of the in-between. It is the ultimate response to the river's question of how to bring the river and its lands to you, the reader, before the river's book launches into the spots that have demanded their own separate chapters, *Bedrock, River's crossings, Regeneration*, and *Litter*.

References

Anzaldúa, G. (1987). *Borderlands/La Frontera: The New Mestiza*. San Francisco: Spinsters/ Aunt Lute.

Nordstrom, M. (2017). (Becoming-with) water as data. In M. Koro-Ljungberg, T. Loyonten, and M. Tesar (eds), *Disrupting Data in Qualitative Inquiry: Entanglements with the Post-Critical and Post-Anthropocentric*. New York: Peter Lang, pp. 35–48.

6 Bedrock's sacramental becomings

Introduction

Bedrock's sacramental becomings is a wondering about how our rivers become sacred and what the elements of their sacramental becomings are, in response to the Ganges' question of how our rivers become sacred. It is the story of a long slow time of illness and healing when I reached rock bottom, the bedrock of my being. Every day at dawn and at midday I was compelled to go to the closest place on the river, marked by its dense row of casuarinas. In the warm seasons I slide in and immerse my body, paddling up and down, up and down, and when the water becomes too cold I sit close to the river and become lost in the reverie of its slow flow. During this long slow time of healing, this place on the river became *Bedrock*, the space of a river's sacred becomings, recorded in river texts, still images and short videos. In coming to this writing, Bedrock's sacred space expands infinitesimally as I contemplate its air, light, sounds, smells, water, rock and stone, and all of its creatures who dwell there with me, all emergent and changing, moment by moment, season by season. I borrow the still frame of Bedrock's casuarinas to create a symbolic holding space for Bedrock's sacramental elements of *Rock and stone; Water; Silver and gold; Blue sky; Air, as the space in between; Air and birdsong; Becoming animal, becoming imperceptible; The spectral*; and *Bedrock's sacramental birds.*

The river's sacramental

I do not know why Bedrock has given me this word *Sacramental*, that has just arrived with the river. So I turn to Google and the dictionary, and find that 'a Sacramental' is a material object, a thing or action (*sacramentalia*) set apart or blessed to manifest the respect due to the Sacraments, and so to excite pious thoughts and to increase devotion to God. Holy water is a sacramental that believers use to recall their baptism; other common sacramentals include blessed candles; blessed palms; blessed ashes; a cross necklace; as well as Christian art, especially a crucifix or cross. I learn that the *Sacraments* have a visible and invisible reality, a reality open to all the human senses, grasped in its God-given depths with the eyes of faith.

Sacramental: I become very uneasy about *sacramental* and *sacramentalia* because of their strong origins in Christianity. I have long felt alienated from the Christian faith, because of what I experience as the violence at its core, and its co-evolution from an earth so very different from this old dry Indigenous land. I talk these terms to the riverlands where I have a river epiphany. Sacramental is exactly the word I want, it is the sacramental of all rivers of the world. It is the multiple materialities of this very particular local river that have become sacramental, the core of a profound eco-philosophical basis of being, knowing, and doing, in this one very precious gift of life. It is silver and gold, water, rock and blue sky, and the space in between, the imperceptible lizard, the spectral, and the swishing of casuarinas in the wind. It is Bedrock. So I ask, how does the river constitute its sacramental at Bedrock?

Bedrock: An extended reverie

Riverlands, 20 September 2017, 6.50 a.m.
Dark silhouette
of casuarinas
frame rising mist's light show
on stage of river's
smooth surface
an upside-down world
of palest blue with pink clouds
dark reflection of trees
all shimmering and
emerging moment by moment
as mist rises and flows in shades
of gold pink mauve apricot
silently amidst twittering
of tiny unseen birds
and beeping of distant trucks
at the dawning of a new day.

During that long, slow, late/summer-autumn-winter of healing, I become with Bedrock's sublime, it is my rock(y)bottom, my Bedrock. I am compelled to the dissolution of self at Bedrock, becoming other with water, rock and (blue) sky. Every day at first light sleepy body blinks into early morning, and again in the bright sun of high noon. Stroll past the tall sentinel trees, spotted gum eucalypts with palest of pale pink smooth skin, reaching to blue sky, wander down the hill and along a dirt track until I reach the row of casuarinas framing a slow sleepy river. Slide into the river where worn dry earth meets river's water, immersing body, dissolving boundaries of skin in water.

Cold of water draws long deep breath as body cruises, amphibian-like, hands feeling along rocky edges of large crinkled rock that leads to overhanging branch of casuarina. The long low overhanging branch marks the turning place, the place where body turns to face downstream, and trawl with the river's flow lulled into reverie by immersion in rock stone water and blue of sky. Up and down, up and down, I cruise as whole being lapses into reverie, boundaries of skin dissolve, lost in immersion, at rest, at ease, at peace, free of pain, at home in the world. It is only here in the river, becoming with light, water, rock, stone and sky that I feel this ancient peace. I wonder if I am going to swim my way through this illness to the other side. Is there another side? Where is it?

> What if it takes sensing the abyss, the edges of the limits of "inclusion" and "exclusion" before the binary of inside/outside, inclusion/exclusion, mattering/not-mattering can be seriously troubled? What if it is only in facing the inhuman – the indeterminate non/being non/becoming of mattering and not mattering – that an ethics committed to the rupture of indifference can arise?
>
> (Barad, 2012, 217)

Sacramental elements

Bedrock's rock and stone

Bedrock, every day, summer–autumn 2018
It's always rock and stone as
hands feel their way
crinkled edges of large rock
place to turn and follow
every rock, stone, crevice, smooth, and crinkled surfaces
large flat rocks round ones tiny pebbles
until rocks and stones meet body
in shallow water of tiny inlet
body turns trawling upstream,
hands feeling their way
to slippy slopy rock of sliding in
back and forth back and forth
hands feel their way taking solace
from Bedrock's rock and stone.

How is it, I wonder, that Bedrock's rock and stone, the river's rocky bottom, becomes sacramental? The sensation of rock on skin as my hands feel their way along the rocky bottom displaces thought, it creates a new sensory language of rock and stone. Rock and stone create a story through my skin that is older than time. Body and skin come into being with the ancient sandstone formed when igneous rocks were produced from the volcanic eruptions of the Earth's still molten core, millions of years ago, the bedrock of the Sydney Basin. The elements of wind and rain wore down these rocks over millennia, carrying the sediments into the seas and back onto the land over and over again. It is the repetition of the refrain that I long for, hands touching this old, old story of the becoming of the Earth. The comforting rhythm of over and over again, coming to know these rocks that are older than time. They are the rocks that hold my body in place against the current, the same sandstone that formed the Bedrock of the Sydney basin that becomes the sacramental bedrock for me each time I swim. '*Touch is* entanglement, sensation, immersion, visual hapticity, ciliated sense, the synesthetic force of perceiving and feeling, contact, affective ecology, involution, strange and wonderful intimacies, sensory attunement, arousal, response, interspecies signaling, affectively charged multisensory dance' (ibid., 207).

Through the permeable boundaries of my body, the touch of skin on rock, I become with this ancient story in that long slow time of healing. It is Bedrock's sacramental rock and stone, making words, making story, making writing, making reading.

A stone is not a rock

For a billion years
the trees have laughed at the wind.
And the sea changed rock
into aging sand
Someday we will be
what we have always been:
Stones in the pregnant bellies of larger rock.
Waiting for birth.
Waiting for the one who
places us gently
into the perfect wave of the wall.

It was rock and stone that led to my first reading after my illness. For a long time I was unable to read, the words on the page swam into my brain chaotically, making no sense, needing rest. Then one day, grieving this loss, I was flicking through the pictures in a book on stone-scaping, thinking about making a stone installation in the front garden when I saw a poem, 'A stone is not a rock'. The words of this poem created such a powerful affective response that I was impelled to read the words out loud, feeling them in my body: 'So much happens in a touch: an infinity of others – other beings, other spaces, other times – are aroused' (ibid., 206).

Bedrock's watery sacramental

Black swans cruise effortlessly
cormorant on branch
twittering all round
head of spectral just visible
why do I need to be here
gathering up this writing
that flows like water
writing that longs to be water
writing that is below the surface that moves

ever so silently becoming water
bits of me going out
and coming in flowing by
in molecular becomings
wibbles like bits of language
floating on river's slow current.

Even before rock and stone, there was water, in the geological beginnings of life on Earth. Water's sacramental is when skin becomes porous, body boundaries dissolve, body becoming water, slide down the slippery rock, roll body over until fully immersed, gliding, not swimming, just floating, hands moving step by step along rocky bottom. Body dissolves in the water, no longer a body but part of the watery river, all parts of self merged with water. Up and down, up and down, watery body cruises, swimming in the amniotic fluid's beginnings, dissolved in the cascade of protons, lost in the geological becomings with water at the beginning of life on Earth:

> How do particles sense one another? Through direct contact, an ether, action-at-a-distance forces, fields, the exchange of virtual particles? What does the exchange of energy entail? How is a change in motion effected? What is pressure? What is temperature? What are the different kinds of forces that particles experience?
>
> (ibid., 208)

Water's sacramental is mostly written in the longing, the cold times when I cannot swim. It is at these times I write of my longing, sitting, back against the rough crinkled back of a casuarina at the water's edge beside the rock where I long to slide in. Today's river is sombre dark green, sun peers through top of trees on opposite bank, now gone behind a cloud. Cold air eddies around my shoulders. No swim this morning, too cold at 11 degrees, body still drowsy from sleep, longs for wake up immersion in river, just longs. It's what makes me part of river through this long illness, the only place where I feel at peace, part of river's slow-moving timelessness with others who are here with me. Water birds offer solace in these times. A cormorant sits on a log mid-river, like me, just sitting, occasionally turning its head, revealing flicker of white face. Swallows flit and flicker, playfully disturbing the smooth surface of the water, and body skin longs to disturb river's still surface with that slide into water.

I find peace lying close to the river's surface, sun warm on skin, not in the water but on the same plane, part of its stillness, almost no flow. I become with the slow-moving flow and its random flotsam and jetsam passing by. A balloon vine pod sails past, pieces of river weed linger along the surface, bubbles congregate in little florets, flowers come adrift from upstream, bringing bright yellow and orangey red to the passing parade. All of us just going with the flow. Sometimes gentle ripples rise up to the still surface from below, wibbles on the surface like the river's texts, a language that emerges from somewhere inexplicable under the water. It is the way I come to be with the river's quiet flow, so still and silent with its random assortment of bits and pieces floating by. 'What is being called into question here is the very nature of the "self," and in terms of not just being but also time. That is, in an important sense, the self is dispersed/diffracted through time and being' (ibid., 213).

Bedrock's silver and gold

> ***Bedrock's every day***
> Silver sparkles of sun
> on dawn's misty river and
> silver ripples on blue
> of bright sunny day
> silver shining on green river
> under cloudy sky and
> silver ripples sparkle
> on black river in the dark
> of evening as sun disappears
> behind dark outline of trees.

Silver is Bedrock's every day. The water's gentle movement over Bedrock's river stones creates ripples that sparkle silver in the patterns of river stones that sit in the middle of the river. Silver ripples flow from its rocks as ripples are lit by the play of light, water, air and movement. There is no sound from the river's slow flow at this spot, just ripples of silver. Bedrock sparkles with silver on pink in the early mists of dawn, on blue in the middle of a sunny day, on green when cloud covers the sky, and on black as night-time comes. Silver is Bedrock's refrain, the song of lights play with water and rock. It is the sublime of repetition that I long for, the comfort of the refrain, the familiar that I desire, the sacra-elemental becomings with silver and gold.

Bedrock offers its first golden sacrament of light's play with rock, stone and water in the middle of that long slow winter of illness. It is a time of no hope of ever reaching the other side. It is early afternoon as I reluctantly haul my body out of the water by hanging onto low branches that lean out over a large rock. Out on the bank the rocky bottom, my bedrock, suddenly shines gold. I stand by the shaft of sun light that shines through the trees onto Bedrock's rocky bottom and photograph a perfect image of shining gold. Every day after that in yoga nidra, the shining of gold on rock presents itself, the longed-for place of ease, peace, rest, as a spiritual home. After this, the still golden light shines over and over again and becomes something like silver's refrain, the ever present sacra-elemental of silver and gold.

Then, one day, gold breaks through the refrain. I am making my way along Bedrock's rocky bottom, hands feeling rocks and crevices, always the same, repetition over time, time-layered, the beginning present at the ending, every moment in one. Then the refrain is deterritorialised by moments of gold, as body, and world become anew in these sacramental moments. This new image of gold is a phenomenon I have never seen before. Beneath me as I swim, golden lines move and merge in glorious patterns on its rocky bottom. Stopping, they become still. It is the waves produced by my moving body that create the intersecting patterns of gold. As the waves riffle the water's surface, the light plays with water and rock, in its infinite dance of movement and change.

Out of the water the intersecting gold lines are still and I photograph the still gold again, but I desire the moving intersecting patterns created by the waves of my swimming body. I'm out of the water now and I can't go back in with my iPhone, even shining gold will not save it, so I walk to the rocky bottom at the tiny inlet where I usually turn, careful not to fall on slippery river stones. I step into the water and the magic begins, and I move my feet back and forth, over and over again, captivated by the magic of that moment of moving intersecting lines of gold created by water's sacramental dance in the space between sunlight, water and rock:

> Not only did the infinities persist when quantum field theory tried to resolve the problem, they multiplied. Indeed, infinities are now accepted as an integral part of the theory: marks of self-interaction—the trace of the inseparability of particle and void. Specifically, the electron's self-energy takes the form of an electron exchanging a virtual photon (the quantum of the electromagnetic field) with itself.
>
> (Barad, 2010, 212)

Blue sky and air

Blue sky, 7 August 2018
Water cool on skin
glance up to blue of sky
river's breath flows
immersed under blue of sky
swim up and down
feeling water rock and stone
until screech of cockies on distant ridge
pulls my whole being
towards distant blue
of sky's space of infinitude
that goes on and on forever and ever.

What is it that makes the sky blue, I wonder, and why does it affect me so in its elemental inseparability from water, rock and stone? There are many different answers to this perennial question of why the sky is blue: Sunlight reaches the Earth's atmosphere and is scattered in all directions by all the gases and particles in the air. Blue is scattered more than other colours because it travels as shorter, smaller waves. This is why we see a blue sky most of the time.

The sky is just our atmosphere as we see it from underneath. The majority of our atmosphere extends about 16 km upward, and this is where most of the Rayleigh scattering happens.

Sunlight, which appears white to the human eye, is a mixture of all the colours of the rainbow. Sunlight can be thought of as an electromagnetic wave that causes the charged particles (electrons and protons) inside air molecules to oscillate up and down as the sunlight passes through the atmosphere. When this happens, the oscillating charges produce electromagnetic radiation at the same frequency as the incoming sunlight, but spread over all different directions:

> Being in touch with the infinite in/determinacy at the heart of matter, the abundance of nothingness, the infinitude of the void that is threaded in, through, and around all spacetimemattering opens up the possibility of hearing the murmurings, the muted cries, the speaking silence of justice-to-come.
>
> (Barad, 2012, 216)

The blue component of the spectrum of visible light has shorter wavelengths and higher frequencies than the red component. Thus, as sunlight of all colours passes through air, the blue part causes charged particles to oscillate faster than does the red part. The faster the oscillation, the more scattered light is produced, so blue is scattered more strongly than red.

Blue light is scattered in all directions by the tiny molecules of air in Earth's atmosphere. The light we see is just one tiny bit of all the kinds of light energy beaming around the universe. Like energy passing through the ocean, light energy travels in waves, too. Some light travels in short, choppy waves. Other light travels in long, lazy waves. Blue light waves are shorter than red light waves. There are other explanations than these scientific ones that come from a very different understanding.

Daniel's grandmother has other answers to the question of blue when she is invited into a school in remote Papua New Guinea to give the children a lesson. She tells them that she would like to teach them about Africa, but she does not know about Africa, so she will teach them about blue:

> 'I will tell you everything I know about the colour blue.' And so we heard about the colour blue. 'Blue is the colour of the Pacific. It is the air we breathe. Blue is the gap in the air of all things, such as the palms and iron roofs. But for blue we would not see the fruit bats. Thank you, God, for giving us the colour blue.' … ' It is surprising where the colour blue pops up,' continued Daniel's grandmother. 'Look and ye shall find. You can find blue squinting up in the cracks of the wharf at Kieta. And you know what it is trying to do? It is trying to get at the stinking fish guts, to take them back home. If blue was an animal or plant or bird, it would be a seagull. It gets its sticky beak into everything.'… 'Blue also has magical powers,' she said. 'You watch a reef and tell me if I am lying. Blue crashes

> onto a reef and what colour does it release? It releases white! Now how does it do that?' … 'A final thing, children, and then I will let you go. Blue belongs to the sky and cannot be nicked, which is why the missionaries stuck blue in the windows of the first churches they built here on the island.'
>
> (Jones, 2006)

Blue is another one of the sacramental elements of Bedrock. Blue is the colour of slow-travelling light waves in the Earth's atmosphere. Blue is Daniel's grandmother's magic phenomenon of the in-between, the gap in the air of all things. Blue entangles ontologies of the old and the new. Blue is the space of the air in between that emerges as another sacramental element of Bedrock.

Air: the space in-between

Rising mist intensifies
looming across water's still surface
as sun remains hidden
behind cloud
crows lonely caw echoes
across river valley from far
white egret flies along river highway
with deep guttural call.
Body longs for immersion
to be part of river
but with no sun it's way too cold
even sitting against old crinkled trunk
cold air eddies around left shoulder
flowing from upstream
in early morning cold
as breath deepens, pausing
momentarily to gaze
at mysterious ripples appearing on still surface.

In Bedrock's ever-changing world, the quality of air changes from moment to moment with time of day, season of year, and the weather with its breezes that flow from the river bringing the river's dank breath into mine. Sometimes at Bedrock the air is dry and smells of dust. At other times, it's a water-wet world with a tinkle of rain on the river's still surface, and droplets of water that shine on the low casuarina branches. I breathe in deeply from the soft moisture-filled air. In summer, the air is hot and it zings with the thrumming of cicada wings. Hot breezes blow along the river ringing with cicada song.

Then the season changes, and hot cicada-thrumming air becomes cold. Walking to Bedrock in the change of seasons is like swimming in currents of air, cold and frosty, then soft and warm, skin swirling in alternating currents of heat and cold. The cold season is the time of mists rising from the river, the meeting of the elements when cold air comes into contact with the warmer surface of water. As the sun hits the water at sunrise, droplets of moisture form, rising and falling, filling the air of Bedrock with flowing mists:

> The elements are an integral force in the production of more-than-human, non-individuated subjectivities. Elements and humans are connected as a pliable matter that bends and folds as though in a sheet that can be stretched and re-moulded. It is a world that exists only in its representatives. It is a lapping of waves, a rumor, a fog, or a mass of dancing particles of dust that enact the vitality of the world. The chaos of the world expresses itself through elements, which in turn affect, enfold and refold into other expressions.
>
> (Piotrowski, 2019, 3)

In the midst of the ever-changing qualities of air, there is one constant, it is the bird song at Bedrock. It is always sound of birds that fills the air all around, greeting as I arrive, continuing through my swim and while I sit contemplating the water in winter. Bird song is so constant it becomes unremarkable, just part of one's being at Bedrock. But what does this mean for sacramental elements? Is bird song the air, the space in between, is bird song a sacramental element?

Air and bird song

Bedrock, every day, time, season
Whip bird calls crystal-clear
calls and calls again
and again
amidst twitter of invisible birds
and bell birds tinkle
as black swan flies overhead
twittering of unseen tiny birds
crystal-clear whip bird call and response
crow's lonely call
white egret flies with deep
visceral craak craak
long lonely call of crows
echoes profound loneliness.

Bird song enters every single one of Bedrock's river texts, so I gather words and phrases from all the texts and weave them together as they sing the song of Bedrock's air. Bird song is always there; bird song constitutes the air of Bedrock; it is air in between of rock stone and water, and the blue of sky. Bird song is Bedrock's sacramental refrain, the comfort of the familiar, as commonplace and taken for granted as air. The refrain is made of bell birds' tinkle, whip birds' call and response, tiny invisible birds' twitter, elegant white egrets' visceral craak craak, swans honking as they fly overhead, moor hens squealing, ducks quacking. High-pitched calls of migratory shore birds gossiping in their spot downstream. Bird call enters my body just as the sounds of the river do, both sound and touch, some deeply affecting like the long lonely call of crows in the background, echoes of profound loneliness, or the whip birds' call and response high in the casuarinas overhead. So crystal-clear and perfect, carrying me upwards and away, soaring through the air, becoming sound, becoming music, becoming with the art of bird song, becoming *Allegri Miserere, miserere mei deus*, singing to the glory of God in the Sistine Chapel, transcribed by Mozart into perfect airborne music:

> The haunting beauty of birdsong, the provocative performance of erotic display in primates, the attraction of insects to the perfume of plants, are all in excess of mere survival: each attests to the excessiveness of the body and the natural order, their capacity to bring out in each other what surprises, what is of no use but nevertheless attracts and appeals. Each attests to an overabundance of resources beyond the need for mere survival, which is to say, to the capacity of both matter and life to exchange with each other, to enter into becomings which transform each other.
>
> (Grosz, 2008, 16)

Then there is the lizard that breaks through the refrain, deterritorialising Bedrock's space and my being, carrying me away to other worlds, of the imperceptible, the mythological, of dragons, dinosaurs and deep time.

Becoming animal, becoming imperceptible

Bedrock, 1 March 2017
Rough crinkled bark
of dark casuarina branch leans
over river
reaching towards water and its
own misty reflections on
this grey cloud day
a leaning stillness until
something moves
ever so slightly
lizard body draped along
leaning branch
shuffles at another presence
on its grey shadowy day.

It is a lizard, an eastern water dragon lizard at Bedrock that leads to Deleuze's becoming animal, becoming imperceptible, and changes the structure and meaning of the river's book. The lizard appears as a dark outline, as if a protrusion of casuarina's crinkled rough bark where it leans over the river. As I move closer, it becomes a lizard, lying long and close against the branch of the tree with its tail reaching right down to the tree's base. Still separate beings, I inspect its stripy reptilian body, dinosaur crest, and lizard legs clasping the crinkled trunk. Slight shift, lizard registers the presence of the other. Human desiring lizard moves forward ever so slowly as lizard and human become im/perceptible to each other:

> If we imagined the position of a fascinated Self, it was because the multiplicity toward which it leans, stretching to the breaking point, is the continuation of another multiplicity that works it and strains it from the inside. In fact, the self is only a threshold, a door, a becoming between two multiplicities. Each multiplicity is defined by a borderline functioning as Anomalous, but there is a string of borderlines, a continuous line of borderlines (fiber) following which the multiplicity changes. And at each threshold or door, a new pact? A fiber stretches from a human to an animal, from a human or an animal to molecules, from molecules to particles, and so on to the imperceptible.
>
> (Deleuze and Guattari, 1987, 249)

Human steps closer and closer with the most imperceptible of movements, iPhone clasped in hand. Pale, watery, yellow eyeball with black iris is rivetted on human eye as human is pulled in closer and closer by the thread of fascination that connects. Lizard emerges in its fullness of being. Bulbous white reptile head is daubed above and below with faint patches of pale pink and a strange irregular black splotch stretches from watery eye to meet stripy body. Large white spines follow the line of head, becoming black as they merge with dark brown and cream stripy lizard skin all the way along its long reptilian body. Bright red replaces stripes under its upper limbs and along the underside of its body. Its tail is long and thin, dark towards the end where it reaches the ground at the foot of the tree:

> That is the essential point for us: you become-animal only if, by whatever means or elements, you emit corpuscles that enter the relation of movement and rest of the animal particles, or what amounts to the same thing, that enter the zone of proximity of the animal molecule. You become animal only molecularly.
>
> (ibid., 274–275)

Edging even closer, human senses the texture of lizard skin, all the colours, stripes, and splotches of black, everything is crenulated, knobbly, rough and grainy. White lizard face sprouts flat spines like the spinosa that crown its head. Neck is wrinkled folds overlapping with red as an upper limb emerges from just under its head. Upper limb has just one bend around the curve of the tree. Lower limb is bent double, revealing red underbelly as it emerges from lizard body and ends in elongated finger claws draped down the rough bark. Human fingers and hands curl involuntarily as they follow the line of gnarled bent claws, crinkled like the rough barked trunk to which they cling. Lizard skin tree skin human skin merged in that moment:

> [A]ll becomings are already molecular. That is because becoming is not to imitate or identify with something or someone … becoming is to extract particles between which one establishes the relations of movement and rest, speed and slowness that are closest to what one is becoming, and through which one becomes. This is the sense in which becoming is the process of desire. This principle of proximity or approximation is entirely particular and reintroduces no analogy whatsoever. It indicates as rigorously as possible a zone of proximity or copresence of a particle, the movement into which any particle that enters the zone is drawn.
>
> (ibid., 272)

Edging closer, human hand reaches out to lizard skin, hand desiring touch, desiring feel of skin on skin, feel of reptile scales, spinosa, folded limbs, long extended finger/claws, and long reptilian body. Closer and closer and closer, human and lizard, until almost meeting when particles between human and lizard move from imperceptible to perceptible. Lizard body detects prey, emits pheromones, puffs out gaping lower jaw, becoming larger around its wide gaping mouth. Then splashes into the river, safe in its watery world, leaving me standing by the lizard tree, two separate beings again in our own worlds:

> Becomings-animal are neither dreams nor phantasies. They are perfectly real. But which reality is at issue here? For if becoming animal does not consist in playing animal or imitating an animal, it is clear that the human being does not 'really' become an animal any more than the animal 'really' becomes something else. Becoming produces nothing other than itself … What is real is the becoming itself, the block of becoming, not the supposedly fixed terms through which that which becomes passes. Becoming can and should be qualified as becoming-animal even in the absence of a term that would be the animal become. The becoming-animal of the human being is real, even if the animal the human being becomes is not; and the becoming-other of the animal is real, even if that something other it becomes is not.
>
> (ibid., 238)

The eastern water dragon's dreaming

All night long, half-waking, half-sleeping, water dragon invades sleep. In the space between sleeping and waking a water dragon dreaming emerges. Water dragon is inside me coursing through my veins with words and images, Deleuze's molecular becomings. Finally I have to tell myself to go into the dark of night. What is this haunting presence of the eastern water dragon, how did this single encounter give rise to such a flight of becoming-other? What is the water dragon? What are its stories, what words does it gather unto itself in this place of Bedrock? How does the lizard's imperceptible constitute Bedrock's sacramental?

It was 1831 when the 'water dragon species' was first described in Australia. Bedrock's eastern water dragon is of the genus *Intelligama*, an ancient lineage of reptile that has existed for 20 million years. The water dragon's sacramental lies in its deep time history, its species naming linking it to the lineage of dragon, connected to the great dragon mythologies of the world:

> What we are saying is that every animal is fundamentally a band, a pack. That it has pack modes, rather than characteristics, even if further distinctions within these modes are called for. It is at this point that the human being encounters the animal. We do not become animal without a fascination for the pack the multiplicity. So the pack is the imperceptible mechanism of the alliance, an animal peopling, the alliance between animal and people and the animal people of the human.
>
> (Deleuze and Guattari, 1987, 239)

Dragon's artworks

Becoming imperceptible with the water dragon at Bedrock is never a matter of its individual sacramental materialities. It is quintessentially a mythological dragon. Searching for the mythological dragons of the Dreamtime, I learn that they appear in the folklore of many cultures around the world. Dragons appear in western cultures around the High Middle Ages and have mostly been depicted as winged, horned, four-legged, and capable of breathing fire. Dragons in eastern cultures are usually depicted as wingless, four-legged, serpentine creatures with above-average intelligence. Dragon-like creatures are first described in the mythologies of the ancient Near East and appear in ancient Mesopotamian art and literature. Stories about storm gods slaying giant serpents occur throughout nearly all Indo-European and Near Eastern mythologies.

It is the proliferation of artworks that represents the multiplicity of dragon's dreaming creations. There are hundreds of images of extraordinary dragon artworks from all over the world: a pale green dragon with forked tongue of fire from Germany; a carving of paired gold dragons on a wall in Beijing; a blue and gold mosaic of *mushussu*, part-snake, lion and bird from Babylon; a papyrus manuscript of the sun dragon, Ra from Egypt; a black and white drawing of the destruction of the sea dragon Leviathan from the Psalms; an ancient drawing of *Rustam* woven into the branches of a tree from Persia; a Greek pottery vase showing the slaying of the multi-headed serpent; a pen and ink replica of the old Norse legend of *Volsunga* in Sweden. All these and more from England, Ukraine, India, Korea, and Vietnam, and a proliferation of dragons from China in every art material and art modality.

Dragons, dinosaurs and birds

This strange night-time dreaming continues to follow its trail of thought in Bedrock's sacramental becomings as I search for where all of these dragon dreamings came from, how did they emerge in all these places of the Earth's deep time and what does this mean for Bedrock's sacramental, where is it leading me? Stories and images of dragons are thought to have been inspired by ancient discoveries of fossils belonging to dinosaurs. Dinosaurs' fossil remains are present on every continent of the planet offering manifestations of dragon dreamings. Elaborate display structures such as horns or crests are common to all dinosaur groups, and some extinct groups developed skeletal modifications such as bony armour and spines. Many prehistoric dinosaurs were large-bodied and were the largest land animals of all time. The vast extent of some dinosaur groups, as well as their seemingly monstrous and fantastic nature, have ensured dinosaurs' regular appearance in popular culture throughout the ages.

Strangest of all, is the story of the extinction of dinosaurs that takes me right back to Bedrock. The fossil record demonstrates that birds are modern feathered dinosaurs, having evolved from earlier theropods during the late Jurassic Period. They took to the air to survive. Birds are the only dinosaur lineage to survive the Cretaceous-Paleogene extinction event 66 million years ago. Dinosaurs can be divided into *avian dinosaurs*, or birds; and *non-avian dinosaurs*. Birds, at over 10,000 living species, are the most diverse group of vertebrates besides perciform fish.

Bedrock's sacramental birds

> ***Bedrock, 26 June 2019, 3.45 p.m.***
> Heavy grey winter clouds loom
> over dark green river
> *Currawongs calling.*
> White-breasted cormorant sits
> mid-river log
> slow passing of salvinia
> leaf stick weed
> *Currawongs calling.*
> Tiny bird flicking
> hopping rock to rock
> moor hens squeal
> *Currawongs calling.*
> Silent black swans perch on
> fallen tree home
> amidst dozens of small black
> water birds cruising
> as two ducks skitter along water
> flicking as they go
> and still the
> *Currawongs continue calling.*

We inhabit Bedrock together, waterbird and human. There is the cormorant who perches on a log mid-river, oblivious to human presence. Each doing their own thing. Cormorant sits, wings spread out to dry, human trawls up and down, up and down, imperceptible to each other, until human hand drops too low into deep water, splash splash human hands, flap flap cormorant wings, flapping as it flies off down the river. Human and cormorant become perceptible to each other as the pact, the thread that connects, is broken.

Human body sits against broad trunk of casuarina, crouched against the cold, unable to swim. White egret, grey in rising mist, huddles mid-river like an old man in a winter overcoat. Swallows dart, flicking at water's still surface, over and over, playing, tempting human who cannot resist and slips into water, playing alongside swallows. Egret flies in to land near human, walks on long fine legs to riverbank hidden from view, and flies again from hidden place as human gets up to leave. Tiny wren drops vertically from tree to water, then back to tree, flicking and flitting, and human thoughts become bird-like, flitting and flicking with bird. A sea eagle soars silently down the river highway lifting human up up and away, soaring into the heavens on wings of flight. Pelican sails majestically past, way too self-contained to notice human admiring its monastic self-contained ownership of the river place. Crows' long lonely calls echo profound loneliness of human on that day.

Bedrock's swansong

Human walks past a bed of river stones and scrambles through bushes to the spot of swans, a fallen tree that creates a place for water birds to hang out. Human hears the swans, treading lightly so as not to disturb their swan world. *Coo coo coo* two black swans talk love to each other, *coo coo coo* again and again. So soft, so gentle, and loving. Human creeps into bushes nearer to swan spot, crack crackle the bushes go, and swans go *squark squark squark*, high-pitched angry squealing. *Squark squark squark*, this is our spot go away. Human doesn't, she sits quiet and still, taking in the calm of this warm place so close to the river, just near her Bedrock. Soon it is swan love song again, soft gentle *coo coo coo*, lulling all of us into restful ease as the haunting spectral figure calls from across the river.

The spectral

Bedrock, 8 November 2016, 4.15 p.m.
Tall grey figure standing
on the other side
of deep green cloudy day river
leaning towards
bright green spiky grasses
head bent
shoulders dropped
arm draped across upper body

head clothed in bonnet
long garments dropping down
to bank of river stones
soft light brown against
smooth dark green river disturbed only
by slightest silver ripple
and protruding twigs
of fallen branch.
Is there another side
where is it?

I have come to Bedrock from my first meeting with Leanne. It was a long afternoon of deep contemplation in her artist's studio, a lifetime of artwork from stories of the Darug people. There were images of the stolen generation of Aboriginal children, adults imprisoned behind bars, and the powerful image of Maria Lock in her multi-layered ochre-coloured landscapes. Other images, always haunted by the past, resonate with all that was lost to the Darug peoples, and what is recreated in these paintings. They were all with me on this early evening at Bedrock when I came to process all that I had just experienced in Leanne's studio. It was there on this day, at this moment, that the spectral figure appeared, tall, grey, head bent over as if clad in a bonnet, arms folded in front, in a position of repose, long flowing garments that reached to the ground. A woman in colonial dress, a woman of the in-between, the meeting place of the Aboriginal past and the contemporary present. 'Each relationship generated among living and nonliving will always be singular, create a different territory, and, by extension, do different work' (Nordstrom, 2018, 150).

The casuarina's story

I sit with her in the quiet peace of the still slow river, watching as the sun goes down and the birds start to settle for the night when just on dark, a wind blows up and the swishing of the wind in the casuarinas tells Noel Nannup's story of Bedrock's deep time sacramental knowings:

> Every person is born in a catchment, so the rain falls, the water flows, so does spirit and they're all intertwined and as we move across land we shed from largest organ of body, which is our skin, and it falls to the soil and that skin is then taken down and if the tributary is full, then it's flotsam on top of the water till it reaches main tributary, that stream will run into until it banks back and as it banks back, then it settles on both banks and for us as aboriginal people we were dancing and singing one side of that tributary the women, the other side, the men, of course, the skin would be taken down into soil through mycorrhizals and the vascular bundles of the sheoak tree and our DNA would be in the tree.

> Those trees will only grow where there is close surface water and as it does that it means the name we give those trees is kwell, some quellala, the tree of our ancestors, the tree of names, the tree of the spirits, and on a windy day you will hear the voices of every person that's ever lived and their skin has been shed and taken up by tree and you will hear their names … everything was controlled by the elders through that very simple understanding of how important water is. The person who's born in that catchment needs to know where all the water comes from in that catchment and where all the water goes to … That's how important water is to people, it's the giver of life.
>
> We have a story that begins a long time ago and it's not about the planet but the things on the planet and that story we call the carers of everything purely and simply because if you were to say it in language, you would have to say *mundalarkagarrinninni* and I hate to think what people would do with that when they try to say it, but of course at that time it was freezing cold, the land was flat and featureless, there was no wind and it was almost dark and in that freezing cold windless landscape all the spirits that would become real wandered, and as they wandered they began a process of elimination as to who would win the right to be the carers of everything and as they went about that, a great serpent that we call the *woggaa* moved across the land and said to all the others, I've got no hands and I've got no feet, I couldn't possibly care for anything, and as it moved away moving across the land, it created great valleys wherever it went. What happens to water when it's freezing cold, it freezes and what does water do when it's freezing cold, it moves in large masses and as it moved in a serpentile action, it carved out great valleys and pushed up hills on either side and left them there as it moved across the land and as it did that, it knew that one day it would thaw out and that it would sit in great bodies of living water. Our *wakkaal* is not only a spirit it's solid ice, it's not only solid ice, it's fluid as well, and it's in everything.

The casuarinas at Bedrock sing of deep time of creation when 'the land was flat and featureless, there was no wind and it was almost dark and in that freezing cold windless landscape all the spirits that would become real wandered'. It is the time of the sacred beginnings of all of the rivers in their glacial sources, memorialised in the sacred water that emerges from Gomuck with its life-saving bacteriophages. It is the time of Bedrock's healing, the time of creation that is sung over and over again, a holding space for Bedrock's sacramental becomings with rock and stone, water, silver and gold, blue sky, air, bird song, becoming imperceptible, the spectral, and its sacramental birds. It is the eternal recycling images of gold and silver, the silver sparkles on the rocks at all different times of day, the stillness of gold in the sun's play with rock, water and stone, and the moving intersecting gold of a human body's movement in water, integrated into Bedrock's sacra-elemental becomings.

References

Barad, K. (2010). Quantum entanglements and hauntological relations of inheritance: dis/continuities, spacetime enfoldings, and justice-to-come. *Derrida Today*, 3(2), 240–268 doi:10.3366/E1754850010000813.

Barad, K. (2012). On touching and being touched. *Differences: A Journal of Feminist Cultural Studies*, 23(3). doi:10.1215/10407391-1892943.

Deleuze, G. and Guattari, F. (1987). *A Thousand Plateaus: Capitalism and Schizophrenia*, trans. B. Massumi. Minneapolis, MN: Minnesota University Press.

Grosz, E. (2008). *Chaos, Territory, Art: Deleuze and the Framing of the Earth*. New York: Columbia University Press.

Jones, L. (2006). *Mister Pip*. Melbourne, Victoria, Australia: Text Publishing.

Nordstrom, S.N. (2018). Spectral data experiment. In M. Koro-Ljungberg, T. Loyonten, and M. Tesar (eds), *Disrupting Data in Qualitative Inquiry: Entanglements with the Post-Critical and Post-Anthropocentric*. New York: Peter Lang, pp. 151–160.

Piotrowski, M. (2019). 'An atmosphere, an air, a life:' Deleuze, elemental media, and more-than-human environmental subjectification and education. *Environmental Education Research*, https://doi.org/10.1080/13504622.2018.1485134

7 The river's crossings

Introduction

The river's crossings invite the reader into a playful multispecies, multi-elemental, becomings that use the metaphor of the site of the river crossing as a means to 'cross to the other side' of being and knowing. Children become sand birds, rock jagwars, mud girls, seal babies, see a crocodile, and a mermaid, crossing to the other side of the human species with relative ease. Animals insert themselves with equal ease into these stories of human child becomings. Ducks and eels congregate to feed on bread scraps, wallaby and wombat avoid human contact, captured turtle hurries back to water, tortoise stubbornly refuses to move, and crow is ever present in its becoming with human desires. The river's crossings are never singular, this is so powerful in its imaginative potential that it gathers into its own spot a multitude of crossings over to become the metaphorical site for all becomings-other-to-oneself. Humans become animals, animals become human, humans become elemental with mud and water, rock and stone as river becomes infused with multiple multispecies and multi-elemental becomings. It is experimental, playful, and transgressive in its writing, challenging the writer and reader to think themselves into becoming with the life of the other, echoing the Yangtze's question about the salvation of our rivers through the stories of human becoming animal and animal becoming human and more.

River crossing

THE TAKING OF UNAUTHORISED VEHICLES INTO THIS AREA IS PROHIBITED

Big red 4-wheel drive ute
parks at river crossing.
4 guys get out and
begin to decompress tyres.
Me: 'Hello, you can't drive into the river.'
4wd: 'OK', unpack fishing gear picnic food and light a fire.
Me: (So lovely to see people enjoying the river despite vehicle prohibition)
Time passes.

4wd: 'Is there a track around there, someone said there is good fishing?'
Me: 'No, there is no track around there, that is the river.'
Time passes.
4wd: 'Where does that track go?' (pointing towards crossing)
Me: 'There's no track around there, that is the river.'
4wd: 'Thanks'. four guys hop into ute, big 4wd ute
rumbles and crunches across river stones
disappears around little island
into bushes on the other side
and never returns.
What is the other side?

A dirt road embedded with river stones, rutted by water flow as it heads towards the river, leads to the river crossing. At the bottom of the road a row of massive sandstone blocks is placed at the water's edge to prevent four-wheel-drive vehicles from driving across the river to the other side. The vehicles use winches to shift the blocks, the blocks get moved aside, one sits permanently in the river, one along the road. They gouge deep wheel tracks in wet ground to make their way around the blocks and rumble across the bed of river stones and around the little island to get to their desired play place on the other side.

There are many stories of what happens across there. Pig hunting as they head across with pig dogs in cages, are there pigs on the other side? Dope smoking, fishermen say they see bongs strewn around the bushes, some say it is grown there. One story tells of a policeman who was tasked to investigate what happens on the other side and he was shot at and climbed a tree to escape. What did he see, or what was he not supposed to see? No one really knows what happens on the other side, except that it is a transgressive place of adult play, a place of crossing over. The river's crossings equally invite children, dogs and others to play at this spot, a place where everyone can become other to themselves.

Sand birds

River crossing, 16 July 2016, 12 noon
Cold winter's day at river
sun shines on warm grassy patch
at edge of track that
leads to river crossing
a perfect place for winter picnic.
Two ravens land high
in casuarinas overhead
to explore picnic
land nearby on low-
hanging branches.
Raven lands on track beside picnic
glossy black feathers lit by sun

parades along dirt track
Luna follows on all fours being raven
How does a raven walk?

Luna sand bird begins stirring up surface of dirt road with stick to make sand. The simple circular-shaped indent in the sand shaped by Luna's long, delicate, sensory fingers soon becomes three as three sand bird girls play together. Cheri, the youngest, begins by obstructing, then joins in, and then Ami too. Each sand bird girl soon has a smoothed-out indented nest in the sand, large enough to sit in, and then a small nest is added to the side. Three sand bird girls begin scratching, stirring, shaping, storytelling all the while. They are sand birds making their sand nests, laying their eggs (small rounded pebbles), counting their eggs. Making, doing, talking, storying. All together with sand, pebbles, sticks, larger stones, being bird.

As different materials get added, different stories emerge. The small nests are filled with a collection of eggs, small, smooth round pebbles. Soft bright green grass growing along the edges of the dirt road is collected by busy sand birds, and placed sometimes over, sometimes under the eggs. Each time something is added, there is much counting of eggs, and stories about eggs. Sand bird Cheri brings piles of little sticks that become small fires next to each little nest. The sand birds don't sit on their eggs to keep them warm but warm them with little stick fires. Sand bird Cheri brings more sticks, big long sticks this time and they are arranged to form a square frame around their hole-making place to mark out the sand birds' 'territory'.

SAND BIRD LUNA: *If I* (fingers fondling grass) *underneath this grass is my baby eggs.*
SAND BIRD AMI: *See my baby eggs* (picking rocks from pile placing on grass)
SAND BIRD LUNA: *We are so smart, we're just like humans.*
SAND BIRD AMI: *... ah stupid eggs...*
SAND BIRD CHERI: *I have to go outside* (sing sing sing sand bird songs)
SAND BIRD LUNA: *We're sand birds. Birds that live on ... in the sand.*
SAND BIRD LUNA: *Make nests in the sand.*
SAND BIRD LUNA: *Ours eggs are hidden under the grass. We don't sit on them. They hide under the grass and we make fires right next to them so they stay all warm.*
SAND BIRD CHERI: *Mine are so cool.*
SAND BIRD AMI: *Because they already get warm by the fire* (long, upward inflection on 'fire').
SAND BIRD LUNA: *Yeah, cos' we think like humans. We think like humans and here's our collection of herbs we need. We're pretending they're herbs.*
SAND BIRD LUNA: *And I need to put rocks around them so they can't blow away. Won't be able to blow over the rocks. Aren't you going to show us the rockiest place, Cheri?*

Sand birds Cheri Luna Ami run down dirt road.

SAND BIRD LUNA: *I really need it. We really need so many rocks.*

SAND BIRD CHERI: *Good rocks. Oh, we haven't been outside the property? since we made the property?*

SAND BIRD LUNA: *We made the property so no one could attack our eggs. There's a strict rule that we must make our own little wall* (Bird sounds).

(Three sand bird girls running back and forth with rocks): *The boss said that we don't...*

SAND BIRD LUNA: *No, we don't have a boss. It's just the way we work.* (Sounds of rocks clinking together as they are placed in a pile on the ground). *It's just the way all sand birds work, it's just what they do, they make walls around their houses, so perhaps they share walls with other people.* (Sounds of girl sand birds). *And that's what we're doing cos' we share in the place* (all jumping, running, making girl-sand-bird sounds). *We all share this area cos' we all like this spot.*

The sand bird game continues with sticks, rocks, and grass endlessly rearranged until a long spiral of larger grey rocks ends in a circle of bright green grass with a carefully placed arrangement of small egg-stones. All is still enclosed by the outer square of sticks, and surrounded by other sand bird nests ...

The girl-sand-birds decamp to another sandy spot on the edge of the road where they stir up dirt again, then make a T in stones. After a little while they return to their sand birds' nests and take up their play again to bring it to its rightful conclusion. Sand bird Cheri leaves first. Sand bird Ami completes her circles of larger stones around her nest. Sand bird Luna is busy counting her eggs and working out aloud how they can only lay 20 and the rest have to go to 'the egg orphan school'. Then a further problem arises with the eggs, what if they laid 30, then how many would have to go to the orphan school? Counting on her sand bird fingers, she tweets, it's 10, so 10 of my eggs will go to the orphan place.

The end.

The sand birds' play is done, now the turtle wants to have its run.

Turtle's story

I'm a happy turtle in my home at the river crossing in a big open pool where the river is slow, like the turtle. I'm always safe there as long as I don't go near the crossing stones where the cars roar across, even my hard shell would not save me there. So I stay safely in my pond. Ducks swim by, egrets visit, cormorants, moor hens, sometimes a pelican, I know them all. Tiny fish always swim with me, sometimes I even eat them. Mostly there's plenty to eat in the reeds by the riverbank, insects, snails and worms.

One day there was nothing. I was feeling so hungry and right in front of me, there was a dangling worm that I love to eat. How good is that to find a worm right in front of me? So I grab the worm, then, ouch, something pricks into my mouth, ouch, and, whoosh, I'm hauled right out of my pond. Dangling on the end of a fishing line. I've seen them before but never in my mouth with my neck sticking out unable to pull it back into my shell. Nothing will save me now. A human hand feels into the soft parts of my throat and, ouch, more ouch, levers out the thing that is sticking into me. The human hand keeps holding me, turning me over and over.

The human hand puts me softly down onto my riverbank. But I don't know what to do, where am I?, what do I do next? A tiny human touches my soft neck and I just leave it out waiting to know. The tiny human goes oooh aaaah, then sits beside me on my riverbank. I can't stay there, it's not my home. Pheew, at last I smell the water nearby, my legs run as fast as a turtle can, all the way to the water, and plop, plop, I am safe in my pond again, living happily ever after until one day I see more tiny humans, but they only play with the mud and I watch them invisibly as they go back and forth, back and forth in their game.

Mud girls

Mud girls 1# Mud kitchen

Silty brown chocolate brown river
white plastic bag floating by
one girl on edge of river in mud
another leaning
over sandstone block
places upright stick in mud cake
dog shake held in suspension
until video starts
and all becomes motion.

The river has turned a rich bright brown, leaving layers of mud as the water reclaims some of its floodplain after an east coast low dumps metres of rain. No picnic today, two girls, Luna and Cheri desire only mud. Luna in mud at very edge of water leans down rolling hands along mud, gathers up handfuls of mud, hands opening and closing, lifting, dropping, lifting dropping, handfuls of mud. Luna looks up briefly from leaning absorption-in-mud-action '*we need more dirt to make more mud in our mud kitchen*' and stands shaping a mud ball with both hands and carries it to the mud kitchen, placing the mud ball onto her mud cake.

At the same time. Cheri is patting and shaping another mud cake on the same sandstone block kitchen, leaning down, placing a stick upright in her mud cake. She stands up from her leaning position, looks around, '*We just need some privacy.*' Luna moves to place more mud and continues Cheri's sentence, '*And we need more water to make it mud.*' Cheri turns and swivels around in a single movement towards the river, enters the river, leans over towards the water, gathers water in her cupped hands, stands and flings water, then flicks water towards mud with feet, over and over.

Luna continues, '*To make the bits that are dry. This bit is very dry*', and as she stands, twirls body around three times pirouetting on one foot at the edge of the water in mud, '*Not muddy enough.*' She leans down to the mud at the edge of the water, swishes hands in large circle movements with extended fingers, making patterns in mud: '*Need to get mud and mix it with not-muddy.*' Cheri meanwhile is facing the river, leans over into shallow river water, legs astride, pushing/lifting/throwing water with her hands back and forth through her legs towards mud '*There. There's some good water.*'

The large sandstone blocks placed by the Council at the river crossing to stop cars driving through the river serve as their mud kitchen, a platform for their mud cakes. A stick makes a candle, but it is mostly immersion in the physical sensory pleasure and qualities of mud and water. The idea of human makings with mud in the mud kitchen is being displaced by the more compelling questions posed by mud and water at their interface. What is water and what is mud, how is water and how is mud, as the girls become water/river, swishing water from the river to make dry mud wet and gather up mud to form shapes, making wet mud dry?

Mud girls #2: Mud monsters

Muddy edge of river
sandstone block covered
in mud
handprint in mud on block
block partly obscured by child body
squatting
feet braced hands down flat in mud
another to side of frame
half visible
with handful of mud.

Sound of river swishing is louder. Luna is squatting in mud, lifts up hands holding two handfuls mud, smiling at them, '*mickles*', leans down and pushes forward into mud with both hands, hands and body lunge forward, pushing mud in direction of water's edge. Cheri also swings into action, squeezing mud tightly with both hands, making loud squeezing sounds, turns and walks into water, leans over, swishes hands clean in water, turns listening to Luna. Luna is vigorously picking up mud with two hands and flinging it towards water while storytelling, '*and then some humans, um, was trying to get the mud away … get the mud to go away, see was throwing the mud away*'. Leans forward again, picks up more large handfuls of mud, throws with both hands sideways towards river. '*They were throwing all the chunky bits away.*' Sounds of mud splash as it lands into water.

Cheri walks towards the sandstone block, leans over to the mud, replies to Luna, '*But then it landed splash*', as she leans over and picks up a handful of mud, drizzles mud slowly through her fingers, then taps her foot on the spot where the mud has dropped. Luna stands and throws more mud into the water, crash fling fling, '*and then I accidentally dropped one on one of the mud monsters swimming*'. Luna picks up more mud and throws it into the river, hands and lower arms covered in mud: *Well, that wasn't good, it rolled over when it hit it. That's a lot of mud though.*' Splash again.

Cheri climbs onto the sandstone block on all fours, feet and hands flat on the mud on its surface, climbs down, and, picking up mud, plops it onto the mud surface of the sandstone block, '*What about we call rocks, mud piles?*' Luna turns back from the river, mud in both hands, '*No, I'm just swimming*', leaning over, flinging mud between her legs. Cheri is walking back and forth to the mud-covered sandstone block, each hand full of mud, '*and then, it crash landed accidentally*'. Luna leans over, hands in water, her back towards the river, legs astride at the very edge of the mud and the water, scooping up handfuls of mud and flinging them into the river. As Cheri speaks, Luna stands up with mud in both hands, drizzling down clothes, dangling in the ends of her long hair, *O.M.G.* Game over, Luna and Cheri stand up, facing each other in muddy bodies, in the place of mud.

The action of the second game takes place in the very restricted space between the sandstone block, the mud and the river's edge with Luna and Cheri so close to each other and so in sync, it is hard to tell who is speaking. Actions and words are so intertwined, spontaneously erupting from the irresistible sensory feeling of mud. Sliding hands along the mud, scooping up mud and flinging it into the water. Mud oozing through toes and fingers. Mud everywhere.

The mud monsters appear unexpectedly in a return to a river crossing story of three years previously when the monster cave was a feature of every walk. On one occasion, Luna and Cheri ventured very nervously into the monster cave, a place inside the low-hanging branches of a tree growing over a storm water drain. After the rain when the storm water flows down from the ridge, the monsters can be heard, gurgling and rumbling inside their cave. Tip-toeing in, '*Shhhh, don't disturb the monsters.*' Luna assured Cheri that the monsters had gone down to the river crossing to bathe their babies. There the mother and father monsters would hold their babies under the water so they had to blow bubbles. Soon Luna could hear the monsters returning, '*Quick, quick*', she called out to Cheri, '*Run, run*', the monsters are coming, and they both ran out of the monster cave and back on the track to their home.

When the river crossing is covered with mud after heavy rain brings flooding, the monsters reappear as mud monsters. Luna distinguishes herself as 'human' but covered in mud from head to toe, Luna and Cheri are as much mud as human, their actions, sounds, stories, imaginations, and words emerging in simultaneous interactions with mud and the river at the very edge where water and land meet and merge.

Words often fail as all of this play occurs at the limits of language where vocalisations are often indistinguishable; words merge with each other and with other sounds, and words are difficult to find for actions. New words come into being from mud and waters becomings. Actions, gestures, words and sounds are all simultaneous and equally arise in synchronicity with mud and water. It is not the river itself which is the main event, but the liminal space at the very edge of the water. It is this very edge space that attracts and is the sole feature of this play. Temporality and spatiality are changed at the river's crossings where matter and meaning are produced simultaneously with water's and mud's play.

The mud finally goes and with the reappearance of the big wide pool, the ducks return and want to have their turn to interact with humans at the crossing.

Ducks

There are four of us wood ducks, and we like to hang out together in the big open pond. We skitter over the surface of the water, running on the water weeds, diving up and down, dipping out heads into the shallow waters. Splash, splash, splash, we go, happy in our little gang. The humans come sometimes to our pond at the river crossing. We flick and splash away, dipping in and out of the water, floating away from them to the other side. Hiding from them. Then one day a small human is standing on the bank, chatting to herself, quack, quack, quack, she goes, just like one of us.

She throws us some bread broken up into little pieces. We cruise over to the bank where we often dabble for food, lowering our beaks into the water and gobble at the bread. Then we leave just in case, cruising to the other side, safe in our big pond. But she is still there, calling us, making those strange little noises, just like we do, so we go again, pushing the water with our webby feet, leaving our trails of ripples behind. She throws us more bread. We dip our heads again, dip and gobble, dip and gobble, until all the more bread is gone. We hang around for a while to see what will happen next. She chortles with happiness and we chortle back, quacking and chatting in a duck-human way, and she chats back to us, chatting and quacking in her human-duck way.

Then the small human flaps her hands in the air, quack, quack, we go. Her hands flap again as she leaves the riverbank. No more bread today. As we cruise off, we know that we will always remember to hang around when the humans come again, in case they have a small human with them who wants to make friends.

Dadda rocks, Mama rocks and bubba kangaroos

Island home, 4 July 2019, 4.29 p.m.
Wade across stones in shallow water
gravelly small at first then big and wobbly
sox in boots wet as always
moor hen calls river gurgles
its ever-present refrain
onto tiny island ringed
with tall casuarinas protecting

> a circle of large river stones
> blackened by fire
> remains of burnt sticks
> from casuarina trees which provide
> this small sanctuary
> half-way to the other side.

The little island sits in the middle of the shallow river crossing, separated by a rocky riverbed from the large sandstone blocks that fail to stop the four-wheel drives from ploughing across to their place on the other side. Children love to play there, it is where Bella saw the mermaid and Ollie the crocodile. For Luna and Cheri too, it is here that anything can happen, they have already crossed to the other side. On this day the water is gurgling its ever-present refrain, the birds are twittering. Luna sits and gathers river stones into a pile while Cheri sings in high bird-like sounds, then walks from the island into the watery pathway of the four-wheel drives' disappearance. She balances on wobbly rocks with a fine stick, then stops and flicks the stick at water and stones, humming and singing along with the gurgle of the water.

Back at the island, the story starts, *that's a daddy* [low sing-song voice, lifting a rock], *that's a daddy, that's a daddy, that's a bigger daddy* [patting a rock each time].

That's a little baby [picking up a small pebble], *that's a little baby, got babies cousins dadda* [arms wide open in expansive gesture]. Walks away lifting arms up towards the sky and crying out a long loud call to the sky. Turns and walks back to rock pile singing

> a-gugu a-gugu a-gugu [sing-song to the rhythm of the birds trilling]. You're a baby [to Luna], I'm a mama kangaroo, I'm a mama kangaroo, you're a baby kangaroo.
>
> That's my fire [loudly, pointing to circle of rocks]. That's my fire, baby kangaroo, that's my fire, baby kangaroo, that's my fire, baby kangaroo.

No human children are present at all in this story, all are transformed to other at the river's crossings. Children become kangaroos, kangaroos become children, and rocks become people, with families, daddas and mamas. The sky is addressed and a new language emerges from the river's song *a gugu a gugu a gugu,* as child becomes bird, singing to the rhythm of the island's bird song. The sky is addressed with calls of wonder. The campfire welcomes them all, the dadda kangaroo, the mama kangaroo, and the baby kangaroo, even though a kangaroo was never seen there, its relative, the swamp wallaby, was spotted briefly at the crossing.

Swamp wallaby

You will hear me lots as I bound through the grasslands, bushes, and trees on my way to the crossing where I drink my fill. I sweep through the bushes from tree to tree, like a passing shadow, from humans I flee. If you do catch sight of me, you will know me because of my dark brown with light yellow to orange on my chest, a light-coloured stripe along my cheek, and tail tip of pure white. I come from a long and ancient line of marsupials with a large family of kangaroos, possums, koalas, and wombats too. I have many human-given names, the Eora people gave me my family name 'wuluba', always forever now called the 'wallaby'. But I am a very special one of all of the wallabies, it is only me who likes to live here, I am swamp wallaby, black wallaby, black pademelon, and black stinker, because I love to live in the swampy riverlands.

I live in the bushlands under the thick casuarina forest of the river's crossing. I hold my head low and tail out straight to sweep in and out of the trees. I forage for food all through the riverlands, feeding on grasses, herbs, ferns, fungi, bushes' leaves and the sap from the golden wattle tree. With my special teeth I can eat the bark and shoots from the needle-leaf casuarina trees. My front legs are short so I can hold the leaves up to my face as I browse in the bushes of the riverlands. I like fresh leaves from the river crossings' bushes best, but if food is scarce I will survive on fungi, bark and algae that grow in the damp shade there.

Humans have observed our extraordinary breeding capacity, so different from human ways. We can grow two babies at once, a nearly born tiny baby growing in one baby compartment, and a newly forming one developing in a second. When born, our tiny undeveloped babies are just the size of a jelly bean. The tiny jelly bean baby crawls all the way through the mother's fur to her pouch, where they suckle the mother's milk. After a couple of months, they're big enough to leave the pouch, but will sometimes jump back in when they're scared.

We keep to ourselves in the riverlands, so you will never see our young, but we are safe there among the rocks, and trees, even when the big cars roll down the road and we hear them from our hiding place at the river crossing.

River camp tattoo parlour and teddies

Waiting, 25 September 2017, 3.30 p.m.
3 teddies, one a furry pinky-beige
with fluffy ears and tail
a second bright orange brown
with pudgy brown nose and black tip
a third light beige
with black beady eyes and nose
waiting
suspended in leafy green bushes.
above 3 girls in summer playsuits
soft bare skin sparkling eyes.
waiting
at model plane place
on road to river crossing

waiting waiting waiting
to proceed to place
of river camp.

Three girls waking at river camp, hair tousled from night's sleep, purple doonas and lavender pillows, crumpled grey tarp and scattered belongings astray around small bodies. Three small bodies poised in focussed attention. Each girl has a texta in hand leaning in towards creamy bare limbs, lips pursed, eyes focussed, hands poised ready for making marks on skin. Luna leans over Ami's calf, drawing in bright green. Ami watches intently as Luna draws, sporting a black bone-like shape along her upper arm, and a green geometric pattern on her opposite thigh. Cheri, lips pursed, frowning in concentration, draws in black on her lower arm, drawing hand already decorated in green and black letters and a leaf pattern on her inner thigh. It's a tattoo parlour, they explain.

River stones tatts

Amidst all the different patterns, logos, and inscriptions appearing on the girls' bodies in the tattoo parlour making, only one tattoo image is repeated. Every girl is adorned with a long coil of green circles twisting back and forth across their lower legs from just above their ankles, expanding into a triangular-shaped bundle of massed green circles just below the knee. They chatter incessantly as they scribe their tattoos, remembering their activities of the day before. '*Remember walking over those stones so slippery and you fell right in*' [laughing]. '*Well, I couldn't help it, it's because it was so scary and slippy where the water washes over those stones*' [more stones on legs]. '*Yeah, but now it's easier cos' we made that trail of rocks where we can walk out to our little swim pool*' [more stones on legs]. '*It was so great when we piled up all those rocks and blocked the water in so we can swim*' [more stones on legs]. '*And then we all got right into the water and it was so cooool.*'

River stones and girls cross over in multitudinous ways at the river camp. They spend much of the day in and out of the river where shallow water flows over a bed of stones, small bodies growing confident and strong as they balance on rocks and create flow channels to divert the water into pools and different flows. River stones make the girls in the place of the river camp. At night they make a circle of larger river stones, a fireplace, where they can gather together around their campfire, making camp, making *ngurra*. Then, when the fire finished, the next morning, they make drawings on the river stones with charcoal from the fireplace, telling stories, making marks as they draw.

There are coiled circles and patterns that mimic the shapes and lines of the rocks, coming to know rocks through their drawing. They also draw themselves and their stories in a perfect transposition of girl and stone. A stick figure of a girl, round face, long plaits flowing from each side, a short stick neck above a triangular-shaped dress, and two stick legs. The letters underneath say her name, *Ami*. And beside this stick figure, a love heart, that simply says *lov*.

The love heart sits between this one and another stick figure. This one has a larger round head, a long rectangular shape from the neck, ending in trousered legs, the letters underneath say Dad, Ami's much-loved father, her love inscribed on river rocks, stones and girl story together as one.

Sticks

Between their tent and the river stone fireplace the girls have constructed their own river camp, humpy with sticks. A fallen lichen-encrusted casuarina branch forms the basis of their structure. Criss-crossing this fallen branch are multiple sticks gathered from all around, thick and thin, twisted and forked, some with dry brown needles still attached, interwoven into a form of shelter. Visible through the structure are their picnic-carrying shopping bags, the blue of an Aldi bag, and the green of a Woolies bag, bright colours against the gentle browns of stick, barks and needles. On top of all this sitting astride one of the sticks, sits the fluffy pinky beige teddy. He's drying, they explained, he got wee-ed on in the night. Girls, body fluids, teddies, flow in this place, temporarily held in place by a fragile structure of sticks, a haven, a home place, made by themselves, for the time they are here at the river camp. I'll tell you a wombat story, I say, a wombat who lives by the river just near where you are camped.

Wombat

Once upon a time there was a wombat here, a short round stubby animals with dark brown shaggy hair and short stumpy legs and this is what she said to me. I'm made for burrowing, my pouch is back to front so that when I burrow my babies don't get filled with dirt. I love to burrow by the river crossing. The earth is rich and deep brown and rises up and down in hillocks, perfect for my home. The river brings down rich brown silt, and the bright green grass I love to eat grows everywhere. Sometimes a nasty vine covers all the grass there, it catches on my short legs and is the only thing to eat but then the humans come to save me. They wear strange bright-coloured jackets, 'fluros', I hear them say. They chop and pull out all the weeds and then they go away. I hide from them, so they do not see me, only my droppings mark my territory.

I belong here in these lands of the Darug peoples. They gave me the many names that I have been known by, 'wambat', 'whombat', 'womat', 'wombach', and 'womback'. I became much loved by all the new humans who came, they wrote story books about me, they put my picture on stamps, and even built sculptures of me in towns. Humans became wombats in shows, in one, they called me 'Fatso' and I got famous in the Olympics. They wrote a book called 'Wombat Stew' for little humans who grow. They named their places after me, 'Wombat State Forest', 'Wombat Hill', 'Wombat' town. Even hotels, in all sorts of strange places, have taken my name. I'm everywhere, man, but I'd rather make my burrow at the river crossings with my humans, turtles, ducks, wallaby, and seals on seal island.

Seal island

Just a little way from the river camp, three girls scramble through bushes to a picnic place a little further upstream. The girls want to play on the rocky island that sits about 20 metres from the riverbank. So we swim to the island, pushing our way through river weeds. At the island the three girls climb up over sharp high rocks that make the shape of an island covered with bushes and casuarina trees and make their way across the island to its other side. On the opposite side of the island, facing away from the picnic spot ,are more high, flat, sharp-pointed, rocks dipping down to the river that become their play place of crossing to the other side.

The rocks are splotched with dried, white, bird droppings, a favourite hanging-out place of water birds. The girls are squatting on the flat rocky surface playing a game that is definitely of 'the other side'. Their father/uncle is out of sight on the shoreline and they understand that anything goes. It is seal island and they are baby seals, alone in the middle of their river world. There are bed rocks and toilet rocks, each girl with their own. Girls are peeing alternately on their toilet rocks, marked with long white strands of bird droppings. They gather bundles of fine green weed at the water's edge, sealbabygirls gathering, gathering, gathering, carrying their weed up to their seal baby home.

They are all seals, baby seals, bleating like baby seals, bleat, bleat, bleat, they go. Soon it is bedtime, the time of songs and stories to lull seal babies to sleep. Luna and Cheri sing lullaby songs in Japanese, lilting girl voices ringing out across the vast empty space of the river. Luna's face opens towards the sky as she forms the Japanese words with her mouth, sitting on her rock beside Cheri who sits on hers, taking her cues from Luna. Ami begins a long story about eels, how the eels come from the water to get the baby seals, the eels are very scary, but the baby seals defeat the eels, sending them back down to the dark places among the weeds at the bottom of the river. Girls' lilting songs and stories drift into the river's air, soon gone, ephemeral as the moment of girl rock baby seal eel and Japanese singing as they are called to return to their picnic lunch.

Ami's eel story

> Once upon a time two eales were out and this little eale ran into the bully's and they started bullying him. So after he got bullied, he swam home and tolld his mum and dad, and they tolld him to bully them back, in the morning, said the little eale. He went out of his underwater cave and said bye and swam to see the bullies and said, hey, bullies, come catch me, and since they're older than the little eale, he can swim faster and then they all stopped and the bullies said, 'whoever can get to the big rock over there first wins.' The little eale said, 'what happens if you's win?' The older eales said, 'if you win, we will be bullyied for life and have to be your servent forever', they said together.

So the race begins 3...2...1 Go! The bigger eales gave the little eale a head start so he zoomed away, in a flash, he was in front of the eales, they caught up to him and then something amazing happened. The two bullies were in frount and then the eale saw them and went in front and crossed the finish line first and the two bullies gasped in surprise.

The next day the bullies met up with the little eale and started the day with a fruty tuty milkshake and started doing all the gob around.

The END

Rock jagwars

The 'boulders' are another favourite place to play, a long row of massive rocks that stretches from the dirt road down to the river. The boulders are an endless source of fascination, some have mysterious holes, others have metal inserts. They tumble against each other, leading down to the river camp place. The rock boulders have been a hairdressing salon, a place for a water picnic, a place of hidden cubby homes in crevices. On this day, they are the home of rock jagwars [*sic*].

Luna sits astride a large lichen rock, upright, narrow beside another similar shape, upright and narrow and in between an installation of sticks. The sticks are bundled across the top, others placed upright through the bundle, Luna leans in, arranging, placing, moving sticks. Climbs onto another boulder, this time fatter with a giant rock beside it, a gaping hole like an open mouth receiving the next bundle of sticks. Cheri hops up and down, up and down, from two more boulders close by, sharp dark boulders, gathering handfuls of casuarina needles from the ground below. '*One, two*', Luna's voice prompts in tune with Cheri's up and down movements.

'*But what if I'm not ready?*', says Cheri as she reaches the top of her boulder and turns facing Luna. '*When I get to three. One, two, three, GO!*' [shouting]. At this prompt, Cheri drops pine needles into the space between the boulders, under the stick installation. '*The hays for our bedding*', says Luna. Luna and Cheri together: '*And every summer we take some off and every cold night we put it on our bed*' [hands moving apart and together]. '*We are rock jagwars, we don't climb on trees, we climb on rocks, our claws aren't very sharp*' [Luna]. '*Our claws is not that sharp*' [Cheri]. The busyness of being rock jagwars continues on and on as girls, rocks, sticks and jagwars together form their ongoing actions and story.

Casuarina's story

Casuarina's secret
I hide my secret in my tree
it's not for you, it's not for me
it's only for those
who want to see
the secret
within the casuarina tree.

Inside my old and mossy
limb
a hollow holds a river stone.
on finest bed
of needle leaves
that holds inside
a hidden box
wherein they write
out of the light
their silent message
to the secret
within the casuarina tree.

These writing hands are human that reach into my tree. They come from some faraway place, to find the secret 'geocache'. They take the stone and then the box and write on pages old and worn, sometimes smudged and sometimes torn. On every page they write their signs, with human dates quite unlike my ancient times:

4/9/13; 15/12/13;
27/1/14; 31/1/14; 7/6/14; 13/8/14; 19/9/14; 31/11/14; 301214;
31/5/15; 20/7/15;
14 Feb 2016; 6-8-16;
2 /1/17; 28/1/17; 18/2/17; 23 April 17;
14 4 18; 19 5 18; 21 7 18; 22 12 18; 27 12 18; 2712 18;
*22 3 19; 6 5 19; 6.7.19.*They write strange names I do not know, strange names like mine was given late. They sign their names beside their dates to let me know who comes to see the secret in the casuarina tree:

Hoogies, Sevabo, Skyejam,
Dolphia & everlasting, footytastics, JFindit,
Willbed 57, ALL7, CHUZE JOSHWAZ7,
Salen and Ros, John D, TAROB 0061,
monkey doodles, midmy7, arwen2,
artemidorus, team merlin dJJCC, Snickerpallan, Bluwassu,
GAZTEK TFTC, Bonfire, Muppetqs, Konie, Nimeg Lovely lady,
MitchPod, Henry Barrett, Martin cousins, BTMU ~ LS m10,
Dogwood 23, old morty, frzzoydaze,
The Neighbs, Trodectus Coibarrous
and Emu Girrl.

Once upon a time a long, long, time ago, our Dharawal people made string for hunting and fishing nets. It was men's work, they used plant fibres, twisting them together by rubbing them on their thighs. It was hard to make string and they thought it was too difficult for the women. But when they handed this work over to the women, all they heard was laughter and chatter. The women kept making the string, they loved sitting in a circle, chatting and sharing gossip. As they worked, pieces of fibre fell to the ground, making a soft mat to sit or sleep on. When the old women died, the Creator Spirit turned them into tall, straight trees, with leaves that looked like pieces of string. If you sit under these trees in my riverlands, the tall straight trees with leaves that look like pieces of string, and you listen very carefully, you can hear the chatter of the old women as they sat making their string in the old days of long ago, the days when emus roamed the riverlands, the emus that gave their name to Emu Girrl, that now lives in the secret hollow of my tree.

Emus

I am maugrai, Emu Girrl, and my family is *ngooree* in the language of the Darug people who dance and sing our stories. They wear our emu feathers, feed from our eggs, always leaving some for us. We are the oldest form of birds in the world, we don't fly but we have feathery wings that look like the grasses where we live. We are the fastest runners of all birds, our shaggy wings flap when we run, keeping us balanced. Our long legs are strong, our heads held craggy and high on long necks and we do not twitter or chirrup, but grunt to each other. We once foraged in the grasslands of the Emu Plains riverlands, searching for fruits, seeds, flowers, shoots, insects, snails, small animals and animal droppings that we grind with river stones in our mouths.

We would trample an area of dry grass in the grasslands to make a platform big enough to hold all of our eggs. When the nest is ready, we lay many large dark-green eggs, maybe five, or even as many as 15. The father emu looks after our eggs that look so beautiful in their blue green against the brown grasses. He will stay there for several weeks without eating because the mother emu is ready to go off and find another mate. But I have to hurry off to breed again, so I leave Maugrai to look after our eggs. Even after the chicks are hatched out, the father emu will keep looking after our babies for two years, teaching them how to find food. I can hear them walking together, my babies softly whistling to each other.

In the time of Dalwah, the casuarina tree, we roamed free in the riverlands of Emu Green, foraging for food in the grasslands, making our nests, rearing our young. We loved to swim in the shallows of the river crossing, flopping into the water, and soaking our feathers before resting under the casuarina tree. Then one day it all came to an end. Explorers saw us and they named their places, Emu Green, Emu Heights and Emu Plains. Their animals roamed on our land and ate our babies, our land was cleared, so we moved to the bushy heights away from the casuarina tree. But Mariuyung, Emu Girrl, still lives here and now she lives in the safe dark hollow of the secret in the casuarina tree.

The other side

So what does the other side mean and where is it? Crossing to the other side invites multiple messy becomings of everything and anything that inserts itself into the becomings that are made possible there. Ravens, river stones, dirt/sand, grass, and sticks become sand bird's nests and territory. A turtle inserts itself into the story when accidentally caught on a fishing line. Small girls become mud after an east coast low floods the river crossing with delicious irresistible mud. Ducks cruise on the big open water, meeting with a tiny human who feeds them bread. Rocks become mamas and dadas on the little island where a mother and baby kangaroo play.

A picture book wombat needs to have its story told as yet another crossing over to the other side. Girls' bodies and river stones become canvases for the drawing of the river's story as teddies dry on stick-built humpies. A swamp wallaby fleetingly inserts itself into the story of the river's crossing, announcing its dark presence briefly just before dusk when small girls tell eel stories and sing Japanese lullabies at seal babies' bedtime. Rock jagwars build lairs of sticks in the boulders that have also been a hairdressing salon and a secret home.

Finally, the casuarina tree makes sense of the geocache hidden inside its broken limb, struggling to understand strange dates and names of those who write inside the secret place of the geocache. Sense can only be made of this through connecting with its own story of the Dharawal women making string who become the forever Dalwah trees in the time when the emus roamed free. The emu story connects the long-ago Dreamtime story of the Dalwah tree with the present naming of Emu Green riverlands. It is through Emu Girrl that the story can be told.

The point of all this is that there is no point, it is an invitation to play, to play at being other to ourselves, birds, animals, sticks, stones, water, mud, and so on, to playfully explore crossing to the other side of being and knowing, in the hope that the life of Princess Baiji's cousin, the finless porpoise, hanging in the balance in the Yangtze River, will be saved.

8 Global materialities

The artful excess of river's litter

With Susan Nordstrom[1]

Introduction

Global materalities dwell in the realm of litter's inanimate manifestations that tell their own stories of movement and flow, stories of the river. They are stories of the inhuman within the human. Litter inscribes its own literacies, some litter with printed text of its own, others with brand names that tell of their multinational affiliations and global flows. Reading litter's textual materialities calls up an exchange with Susan Nordstrom from the Mississippi River in Memphis, Tennessee. Our rivers' waste meets symbolically on remote Henderson Island in the Pacific Ocean with more plastic than anywhere else on Earth. Plastic and waste call us back to our rivers, the Nepean and Mississippi, to think with waste. Waste creates with us, and on us, moves us from its affective production of disgust and aggression, to embrace its proliferation as artful excess. The orange chair sings, waste's installation moves and dances as we struggle to embrace its artful and excessive proliferation in search of a new syntax, a foreign language within the language (Deleuze, 2006). Our thought experiment with waste materialises transformative becomings that generate past-present-future affective residues of wonder about the materialities of litter and rivers. We create with bodies of waste, bodies of rivers, bodies of humans, and bodies of theory.

Walking
struggling with (im)possibilities
of rubbish and litter,
putrification of water,
incursions of four-wheel drives,
mowing of soft grassy bird habitat
until a thousand tiny stars of dewdrops
shimmer in grass as data that glows
could write a whole book about lifefulness
of dewdrops in grass of that early morning walk by the river
each droplet of water made by
precipitation of turning of the world

as day turns to night and night turns to day
each dewdrop clings to its upright blade of grass
diffracting light rays as stars or making
tiny rainbows at my walking feet.
Ephemeral, they are gone
by the time my walk is ending
and yet they hold the whole world
in that microcosm of water.

Litter's story at Emu Green riverlands

Pristine beach and stones

Litter's story begins in 2013 with a small experiment with two small girls to explore Barad's intra-action: how 'the primary ontological unit is not independent objects with independently determinate boundaries and properties but rather phenomena that signify the ontological inseparability of agentially intra-acting components' (Barad, 2007, 23). The girls choose a walk to the river, to the tiny beach that faces the flow of the river to its meeting place with the great Pacific Ocean. The spot is pristine, its only litter is river stones, which they immediately pick up and throw into the water, watching the ripples. Charmaine picks up handfuls of dirt and sprays it over the water where it makes a sprinkling sound as it falls onto the still surface. She then turns to the little riverstone fireplace, 'This is a little fire. It is my little fire, I am making a little fire for *everybody*!' She begins to play in the fireplace with the charcoal and burnt remains of invisibilised rubbish. Lulu smoothes the bumpy surface of sandy dirt with a sweeping motion of one hand, placing flat pebbles in a half circle on the smoothed surface with the other. Charmaine continues her now well-established story of cooking. Lulu gathers up all the flat pebbles from their arranged pattern, throws them into the middle of the little fire and joins the cooking story. 'We're making a camp', she says as I begin to persuade them to move on from their multiple sensory intra-active becomings with river stones, water, sand, dirt charcoal, meaning and matter emerging simultaneously.

Litter's communications

Litter begins to make its presence felt on the next walk to the little beach with the appearance of a cigarette packet, drink cans, plastic bread bag, corn kernel tin, and so on. Curious at first, I decide to explore the litter and for three months I collect the litter every Monday morning. As I collect the litter, a story emerges of silent and invisible conversations mediated by the inhuman. Litter is collected by others, one day there and the next gone, a bag is left full of litter hanging on a tree, waiting, burnt remains of litter in the circle of river stones fireplace, some items of litter are recycled to become something else. Litter tells me who has been there and what they have been doing from what they have left behind, fishing gear and bread bags, large numbers of drink cans, cigarette packets, and so on, varying from week to week. I feel a sense of connection and anticipation in what I will find, how I can come to know the river through its litter. Every Monday I gather it up in Woollies' plastic shopping bags and hang them on the clothesline until all 12 are hanging there.

At the end of three months I photograph each week's collection and document each item in litter's excess of proliferation. Just two examples:

Week 2 – 18 items: 2 plastic bread bags, 2 plastic fishing bags, 2 glad wrap, 1 plastic food container, 1 plastic slurpee, 1 plastic water bottle, 2 tin lids, 1 drink can, 1 drink carton, 1 paper biscuit wrapper, 1 large plastic bag that says 'crickets lge', 1 pce plastic, 1 pce card, 1 tissue.
Week 5 – 17 items: 3 plastic bags 3 x bread, 3 plastic drink bottles, 1 shapes carton, 1 chip pkt, broken glass, plastic lid, plastic fish bag, Edgell's corn can, cigarette butt, cigarette pack, tin lid, paper labels.

The litter inscribes its own story, some litter with images and printed text of its own, and much with brand names that tell of their multinational affiliations and global flows.

Litter's global materialities: bodies without organs

Cigarette packets are the most confronting of these global materialities. The first that catches my attention, a tagged blue foot (dead body?) with superimposed text that says 'Smoking kills'. Confronted by these multiple scatterings of 'bodies-without-organs', I explore litter's inscriptions of its story in more detail, asking how does the in/human of litter constitute a relation with the riverlands and what is the meaning of these global flows of movement and connection mediated by the river?

First, a premature baby, wrinkled, red skin, respirator taped on nose and forehead, disembodied hand cradling head, wrinkled eyes closed, positioned below an empty white bed. White text on black, above these images says, SMOKING HARMS UNBORN BABIES, and below on marketing gold, Bond Street Red. On the other side, the same image of premature baby with taped respirator and hand, no empty bed this time but text that says:

> Smoking during pregnancy reduces blood flow in the placenta and limits the oxygen and nutrients that reach the growing baby. Smoking increases the risk of miscarriage, still birth, premature birth, problems during birth, or the baby having a smaller brain and body. You can quit smoking.

And below, on gold, AUSTRALIAN FIRE RISK STANDARD COMPLIANT/USE CARE IN DISPOSAL. On one side, in black on yellow: 'WARNING/ Tobacco Smoke contains a toxic mix of chemicals that cause disease and early death in children and non-smoking adults exposed to smoke.' On the other side on marketing gold: 'Phillip Morris Ltd./ 30 Convention Centre Place,/ South Wharf Victoria, 3006,/ Australia./ Made in Indonesia./ Phone 1800237246'.

Back, front, top and bottom and each side contain multiple texts and images, designed somewhere by someone to negotiate the tricky terrain of marketing a product that causes death. The cigarette packet embodies these complexities with its bodies without organs, bodies that are, like all bodies, permeable and vulnerable. The toxic chemicals flow through the body in the blood circulation shared by mother and baby. The whole representing an intermingling of bodies with the precise places of litter's textual production: of growing, and harvesting tobacco, turned into cigarettes in Indonesia, and packaged at South Wharf Victoria, to be dispersed to thousands of retail outlets all over Australia. And the strangeness of the disposal warning, use care in disposal, that sits in the ambivalence of the disposal of the dead baby's body, and the cigarette packet itself, which appears disposed and undisposed on the tiny beach at Emu Green.

Then, a human heart shape ,with pale yellow fatty covering, dark red/ brown patches and superimposed text that says: 'DAMAGED HEART MUSCLE RESULT OF CLOGGED ARTERY'. Under that is the brand name, 'Winfield Original Blue + taste flow filter' and the smaller text below restates litter's story: 'Smoking causes heart disease by damaging the blood vessels and blood supply to your heart. Smokers have more heart attacks, repeat heart attacks and angina than non-smokers.' On one of the sides, the text in black on yellow says, 'WARNING: Inhaling tobacco smoke releases hydrogen cyanide into your body, a chemical that is very toxic to humans. It can damage parts of your body including your heart and lungs.' And on the other side: 'Manufactured in Singapore/British American Tobacco. Australia Limited. 166 William St /Woolloomooloo NSW 2011. / Phone 1800 643 562 / 46268C12 0510.'

I think of the veins of the Hawkebury Nepean River that criss-cross the Cumberland Plain of Sydney Basin, and the clogged arteries of the human heart, human and river entangled in the story of the river's litter. The rivers of the world clogged by pollution and dying, and the individual death of the human, when the heart muscle can no longer pump life's essential blood flow through all of its veins and arteries. The essence of flow and movement within and without. The movement of tobacco grown in the soils of Britain and America, travelling to Singapore to be manufactured into cigarettes, travelling to 166 William St, Woolloomooloo, in Sydney, a very precise place with street and phone numbers, to be placed into this packet. This packet of cigarettes ingested into a singular human body, a body that can never be singular as it flows and stops flowing within these global materialities. And finally the (always empty) packet landing on a tiny beach at Emu Green on the Nepean River, in this very local global entanglement of bodies, arteries, flows, and capital.

Litter's continuing global connections

Reading more of litter's textual inscriptions of its own story, I find I am not here but everywhere. A crumpled shiny white can with pale gold stallion rearing above black text shadowed with gold reads 'Nelson County Bourbon', and below on shiny red, '& Cola'. A sleek black can has 'Woodstock' in white, down one side, and across the middle, 'Kentucky blended bourbon & Cola, original blend, 4.8 %'. Another in black with fancy crowded text says, 'Jack Daniel's, Old No 7 brand, quality Tennessee whisky & Cola, whisky distilled by Jack Daniel distillery, Lynchburg, Tennessee, USA'. I find myself not here but in the United States, alongside my collaborator, in Tennessee, Kentucky, and Nelson County. Litter carries me on the flow of global capital to co-river dweller, Susan Nordstrom from the Mississippi River in Memphis, Tennessee.

Henderson Island: our symbolic meeting place

Susan and Margaret meet on remote Henderson Island, a tiny dot in the vast blue of the Pacific Ocean, equidistant between the large continent of the Americas and the smaller island continent of Australia. This tiny remote island is our symbolic meeting place, half-way between our two continents, and the location of more litter than any other surface on the planet. A study (Lavers, cited in Cooper, 2017) found that the island contains an estimated 37.7 million items of debris weighing 17.6 tonnes, the highest density of plastic rubbish anywhere in the world. Most of the items are everyday household goods, such as cigarette lighters, plastic razors, toothbrushes, plastic scoops, and babies' dummies. The litter comes from everywhere. The iconic rivers of the world are all connected through the flows of waste into the river's oceans. From the symbolic meeting place of our river's litter connections, we return to our own rivers, the Mississippi, and the Nepean, to explore our affective river becomings, including with the productions of litter.

Wild thinking with water: dreams and possibilities

Susan and I formalize our exchange of litter's materialities in a proposal for a symposium 'Wild thinking with water' (AERA, 2018). Our paper, 'Meeting of the Waters: Death and the Global Materialities of Litter' is situated within the burgeoning Anthropocene scholarship informed by new materialism (Barad, 2007), staying with the trouble (Haraway, 2008) and climate change's necessity for new figures of human and new concepts of thought (Colebrook, 2010). We connect these philosophical contributions to the move towards post-qualitative inquiry in educational research (Lather and St. Pierre, 2013), from our own 'particularities of cultures, times, and places', stretching our thinking by exchanging images, thoughts and feelings across continents.

Our waters meet and we write with them, not about them. We write with water allowing it to send us through movements over pebbles and rocks. A shared sun moves through our water and diffracts our visions. Spectral images announce themselves in the water and ask that we speak with them. These untimely and incipient images write us as we write them as we work towards a justice to come that is beyond the living present. We write and are written in water's accessible language, images, and movies. We select two examples of uncanny connections in our *Meeting of the Waters*, an affective connection around death and dying with rivers, and a material connection in relation to the global materialities of litter. The second explores the wicked problem of litter, connecting through the litter on Henderson Island, with more plastic than any other surface on earth (Cooper, 2017). We explore dreams and possibilities for thinking within the global materialities of litter from this half-way place between our two continents.

As we write about our rivers, our water entangles 'Margaret' and 'Susan'. Susan says the currents stirred and dissolved us, our waters, ourselves. A flood manifested us into *Emu Girrl* and *Feral Susan*, a way to contingently situate ourselves, a way to care for each other and others in our 'thick, impure, involvement in a world', as we engage in an ethics that is 'a hands-on, ongoing process of re-creation'. In the end, Emu Girrl is unable to attend the conference. Her thick impure involvement in the world leads to burnout, and she becomes part of the burnout society whose medications have entered the flow of our rivers, but Feral Susan gives Emu Girrl the greatest gift of writing in her absence, extracts from which are presented throughout this chapter.

Email 1

> My Dearest Emu Girrl,
>
> I write this email to you listening to the pitter patter of rain falling on my spring your fall day. The soft and rhythmic rain falls on verdant green weeds and tiny purple and white violets and pulls my attention away from this email. The rain makes the branches a deep brown, the branches I refuse to pick up, do you remember that picking up sticks and branches following a storm is my most detested chore. Perhaps I leave them there to see them move between shades of brown, between dry and wet. The seemingly docile white blossoms on the dogwood tree scream out against the grey sky. A small rumble of thunder shifts my attention. An oogy-boogy rumbling perhaps. What you call our connection, oogy boogy, water-sky movements, connections between you and me, here and there, your river and mine. Soft rumbles of thunder during a spring rain that pitter patter into Deleuze's imperceptible. Our oogy boogy. And, I'm reminded of what you said about the language of water. In one of our email flows, you mentioned how you think too of the language of river, you often try to find words to express the different sounds of your river as it seems to have a different sound every time … from silent almost still flow to quiet patter of raindrops, the gurgle as it plays with rocks, a roar of energy in flood, and everything in between. A language of waters meeting and converging together. Oogy boogy pitter patter gurgle roar.

And perhaps that is why I struggled with one of our ideas for this presentation, a selection of our email exchanges that we would read to each other together, our oogy boogy united in time and space rather than flowing across time and space. But, you're not here. Yet you are. The email exchanges become our ever-shifting riverbed from which we flow. Our words become tiny sediments of thought that perhaps only we can say to each other. Words pitter pattering, gurgling, roaring over space and time. Words making our language of oogy boogy. Our oogy boogy becoming imperceptible is always both here and there, yours and mine, present, and absent. And, so I write to you, today, because I cannot imagine someone else reading our oogy boogy, our watery oogy boogy gurgling pitter patter roar language of our connections.

Feral Susan

Litter site mostly clean
except for sparkle of plastic in sun
and empty cigarette box
with traumatic body images
on all sides
made in Indonesia
packaged in Woolloomooloo
circulating flows
of global capital
flowing from somewhere
where the river goes.

Email 2

Dear Feral Susan,

Your letter makes me think of litter's language. Do you remember the story of the orange chairs that appear and disappear all around the riverlands? First, sitting beside each other at the original fishing spot, now the litter spot. One time both appearing on the little island just out from Bedrock and then one was in the river. I hauled it out when I was quite weak from my illness, a powerful physical achievement of balancing on slippery rocks to reach the small island. I remember as I drew close to the island the rocks were decorated with white splodges of bird presences, and tiny birds hopped on the stones exploring the waters between them. Once on the island the orange chair is placed beside its mate, a couplet again. After that they both disappeared, no orange chairs for a long time. Then one appeared at the top of a tree above a dug out and strangely painted tyre, so they travel around seemingly from their own volition.

The orange chair calls to me one day in late winter when I am walking along thinking about our paper proposal for *The Affect of Waste and the Project of Value: The Rejected, The Dross, The Chucked, and/or The Useless.* As I'm wandering and wondering in the riverlands, I hear what I think might be a lyre bird, a high-pitched rhythmic call that I've not heard here before. It's drought and birds are migrating differently, but the only way I can tell it's a lyre bird is by hearing its imitation. You may not know about the lyre bird, but they usually live in the Australian rainforest, and they are known for their song, a mixture of elements of its own song and any number of other mimicked songs and noises. They sing with most intensity in their peak breeding season, from June to August, when they will sing for up to four hours of the day. The lyrebird's syrinx is the most complexly-muscled of all the songbirds, giving the lyrebird its extraordinary ability, unmatched in vocal repertoire and mimicry.

I move in closer to its bushland location, and I can hear what I think is the lyre bird's characteristic call. They are so shy I dare not go any closer and no matter how much I strain my ears, I cannot distinguish the imitations of the lyre bird from the sound of the actual bird calls in the crazy sonic landscape cacophony. Every bird is singing their own song and I need to sit still, silent and patient for a long time so I can hear all possible calls. But where to sit in this wild tangle of bush? Then I remember the orange chairs. The two sitting on the little island near Bedrock, then the one that appeared in the river and then they both moved again. So I go searching for an orange chair, and near the unreachable one up in a tree I find its mate sitting alone on the dirt track. Carry it back through the bushes and weeds to come close to where the lyre bird calls. I carefully place the orange chair in a hidden palace of noxious weeds, balloon vine draped over privet, where I know I will find it again. There I sit, quiet and still on the orange chair, holding out the iPhone to record the sounds, knowing that I will be able to manipulate the sound later, slower, louder, to maybe discern the lyre bird's imitations. As I repeat this over and over during that late winter, orange chair and lyre bird calls inseparably becoming present in these moments of bird song.

Emu Girrl

Sitting long and still
on orange chair ears straining to hear
lyre bird's imitation
from surrounding bird song of late winter
body slows so still can hear
heart beating on ear drums
repetitive black bird call
beat beat beat
switches
to other songs
crystal-clear whip bird calls
twittering of finches
squealing of black cockatoos.
It's bowerbird going right off
enthralled become immobilised
from multifarious singing
of orange chair.

Email 3

Dearest Emu Girrl,

Each time I write to you or see your email appear in my inbox, force expands. An intimacy of flows is generated. Our rivers meet in our exchange and run wild together through encounters that are much like the undulating currents of the Pacific where we presume our mutual rivers meet. Much like the waters of our rivers that carry runoff from farms, the litter that makes its way into waterways and by extension oceans, and sediment of our river bottoms, our thoughts are like the litter of our rivers. They momentarily appear at the surface only to be pulled down by river currents. Our thoughts are these fleeting moments of recognition, moments of life and death flowing together.

So perhaps we write to each other to expand our forces, our flows. Many times your emails come to me when I need them the most, when my water has become stagnant. Your words are rain that animate my stagnant puddles of water and create movements for the water to flow. Water unites us, flows with us, and provides a life-giving force. I long to touch my river and that is what your emails generate for me, a profound longing to not stand above and far away from rivers but to touch them, dissolve into the waters to dissolve binary thinking.

Feral Susan

Bedrock looks like
new litter site
gold cans sprinkled here
there and everywhere
escape to peace
of orange chair
in its palace of noxious weeds
privet draped with curtains
of bright green balloon vine
to record lyre bird calling
amidst chorus of bird song.

Email 4

Dearest Feral Susan,

I am struggling with litter, ricocheting from despair, disgust, anger and sadness, to attachment. I wonder about using your email signature as the framing for our paper: '*A work is supposed to bring out problems and questions in which we find ourselves caught, and not to provide answers. A work is a new syntax, which is much more important than diction. It carves out a foreign language in the language*' (Gilles Deleuze, 2006). I do not have any answers or any clarity of thought, it moves and flows like the water around litter's excess until the orange chair's bird song calls up Grosz's 'art as the excess of natural selection' to offer an entirely new way to think about litter:

There is much 'art' in the natural world, from the moment there is sexual selection, from the moment there are two sexes which attract each other's interest and taste through visual, auditory, olfactory, tactile and gustatory sensations. The haunting beauty of bird-songs, the provocative performance of erotic display in primates, the attraction of insects to the perfume of plants, are all in excess of mere survival: each attests to the excessiveness of the body and the natural order, their capacity to bring out in each other what surprises,

> *what is of no use but nevertheless attracts and appeals. Each attests to an overabundance of resources beyond the need for mere survival, which is to say, to the capacity of both matter and life to exchange with each other, to enter into becomings which transform each other.*
>
> (Grosz, 2008, 17)

I wonder if we can think about litter in this way, do we dare engage in this thought experiment? How can our words encompass litter's embodiment in our bodies? What bodies of theory will work to come to some understanding?

Emu Girrl

Colorbond wildly decorated
two sheets held
on massive timbers
beautiful artworks of multicoloured
greens pinks white orange blues yellow
spray painted symbols giant words
with no meaning
but the identifying languages
of the litter kids.

Email 5

Dearest Emu Girrl,

Our oogy boogy is our exchange, our becoming. For as you wrote to me … The theorist Ronald Bogue (2009) writes that learning and knowing in Deleuze's thinking are what takes place as one organism immerses itself within another element and opens up to an encounter with signs. These signs work like interferences that might disorient or even shock the organism and, writes Bogue (2009, 11): 'forcing thought to deal with experiences that disrupt the common, coordinated functioning of the senses and faculties'. Grosz, just as Bogue, offers us a view on knowing and learning that is about engagements of bodily affects and that has very little to do with the autonomous knower that knows from a distance: Pleasure and pain are the corporeal registrations of the forces of the world, the visceral impact of forces what we use to struggle with and against, in order to become more and other. They are the most powerful aids to learning and the most direct and effective stimuli for action, and thus the expansion of force. (Grosz 2005, 190).

Feral Susan

Over and over colorbond structure
falls back into water
back aches breath comes hard and fast
waste is strong in its resistance
body-becoming strong-with-waste
finally done
stand the bike up push pile of litter into its fold

wash off slimy river sludge and
photograph installation before departing
to new swim hole
soul weeping for rivers of the world.

★★★★★★★★★★

Then there is the other half
wade into water once more,
skidding and sliding over slippery rocks
this piece larger and heavier
back already aching neck sore
but it calls and calls and calls
cannot resist pushing pulling manoeuvring
its awkward water-filled slimy weight
through water and onto bank
even at bank river keeps reclaiming its trash.

Email 6

Dearest Feral Susan,

I have taken your words into my body and am now moving with the flow of litter along new desirelines that take me on another tributary of being and knowing human differently. I have kept on wondering about your question of the litter site, 'have you spoken to the people who litter?' And its spin off with the orange chair that gave me so many bird songs of litter. I was wondering about the multiple languages of litter when last week I listened to a talk by Susan Jacobson from Florida State University about *Creative Environmental Problem-Solving through Art and Ecology Integration.* I was not so much interested in her talk when it turned out to be framed in terms of 'skills' development by linking artists and scientists in higher education (see description below). BUT one of her examples, that was tossed out as a sort of throwaway line, was the creation of a heap of rubbish that people had to walk through as an art installation. The artists documented their embodied responses to the litter. I've been pondering for ages how to create something with the waste from the litter site knowing that my ex-student Sarah Crinall makes her own installations at home with the rubbish she and the girls collect. I don't want to take it home, it seems to belong to the site and I want to resist the idea of being the river's waste housekeeper. SO, the idea of piling up the rubbish tickles my brain and prompts me to go to the river again and becomes a long long story.

What I see amazes me. A tent still erect in situ, and in front, its very own litter installation, including plastic bottles and containers, tin cans, and an outpouring of loose corn kernels from the mouth of very large carp, all of it spray painted orange. This sitting on the charred remnants of a camp fire. There was not much loose rubbish around, so I simply photograph an image of this strange installation. No possibility or space for my own litter creation.

Next day the spectacle is entirely changed. Tent gone, orange remnants dispersed, carp no longer visible, much litter scattered all around. I decide to create my installation. With some difficulty I wade into the edge of the river and haul out a bike, disintegrating in the water and try to place it upright, but the rubbish has its own mind, the bike keeps lying down so I pile the rest of the rubbish in front of it, fishing it out from under the bushes, in the river, plastic bags and bottles, glass bottles, aluminium drink cans, orange spray paint can, and other scraps of everyday life, underpants, metal frame of chair, metal rods, and perfect pair of pink-handled scissors. A small label that I place on top of the pile that reads: The *Original* ENVIRONET *AUSTRALIA*.

But then comes the big body challenge, about 10 metres downstream in the river, the remains of the earlier installation from the site, metal and fibreglass colorbond structure nailed to two long pieces of four by two timbers. It's been sitting in the river for at least a year. I have my swimmers on under overalls, boots and woolly sox, no gloves to protect hands from the harsh metal and scratches of all this creative work, I have red blood streaming down my hands already. I ponder wander and wonder what it might be like, having hauled the orange chair from the middle of the river before with great difficulty. But I cannot resist. I want to add the graffiti colorbond to my own artful creation.

The waste calls and calls to me until I step into the water, gum boots, bubbling and filling, hobbling over rocks with help of one of their fishing sticks. As I reach the structure I lean forward and pull back and part of the structure comes away from the rest, easier, I think, to heave and pull against the resistance of the river. I heave and pull manoeuvring colorbond nailed onto its timber through the water, heave and pull, heave and pull until at the bank the whole comes adrift from the timber but it still won't lift from its watery weight until I see it is firmly attached to another piece of river debris, a small log that has embedded into the metal edge. More heaving and pulling and, hey presto, the log provides a perfect stand for the colorbond screen albeit now covered in river sludge, a backdrop for my installation, just as I imagined.

BUT there is the other half. I ponder, wander and wonder again until the waste calls and I wade into water once more, skidding and sliding over slippery rocks. This piece is large and heavier, my back is already aching, my neck sore, but it calls and calls and calls, I cannot resist the pushing and pulling, manoeuvring its awkward water-filled slimy weight across and through the water onto the bank. I heave the timber up first, this time the colorbond structure does not come away. Each time I pull and push the colorbond end, the timber falls back into the water, heave and push, pull and struggle back aches, breath comes hard and fast from the effort.

The river keeps reclaiming its trash. I am twisted in fishing line, and then a red string tied to a bloody red, half-eaten carp. I unhook myself. Finally, the whole is onto the bank but manoeuvring into position is tricky too, waste is strong in its resistance, my body-becoming strong-with-waste. The metal edge keeps wedging into the soft dry ground so I get a river stone to roll it on and finally with more heaving pushing pulling and rolling, it is done. I cannot make it stand up as I had imagined, so it lies there beside the pile. I stand the bike up now propped nicely on the upright screen, and push the pile of litter into its fold. The bloody red, half-eaten carp is placed in the front of the pile with its string dangling from the bike. I wash off slimy sludgy river and photograph the installation from each direction, before departing to my new swim hole to clean myself off.

The next day I wake, wondering how will my litter installation speak to the occupants of the site? Should I call the council and ask them to remove it, or suffer the anguish of the rubbish going back into the river? What does all this mean about the artful proliferation of the river's waste? I write a note by hand in large capital letters, carefully worded so as not to offend and place the note and a texta inside an A4 plastic envelope of the sort I'm always discarding from student work:

> THIS IS AN INSTALLATION
> ABOUT HOW WE CONNECT TO OUR RIVER.
> IS IT YOU?
> DO NOT REMOVE
> PENRITH CITY COUNCIL WILL TAKE IT
> IF YOU WISH TO COMMENT PLEASE
> LEAVE A NOTE IN THE PAPERS HERE

It's now early morning walk time, a visceral stirring of half-fear, half-excitement, I walk out, it's raining as I approach the site, fearful that all the litter will be back in the river. Walking carrying the plastic sleeve, more anxious than anything else. Who is watching, what will be there?

Heart sinks as I pass through bushes to little beach, massive colorbond sheets are gone, only one large timber remains. Bike is visible now back in the river, some of the rubbish gone (half-eaten fish on red string) but most is still there. More is added to the site in the endless proliferation of litter. I add the additional rubbish and place the bike beside it and photograph the remnants of my installation, changing the *IS* in the sign to *WAS*. Then I explore and find one piece of colorbond in bushes behind so I pull and push again and place it at the back of the pile. I try to stand it up with the timber but it doesn't want to stand, insists on its lying down position. So I photograph and leave somehow polluted and compromised and walk to the new swimming hole, soul weeping for rivers of the world.

I ponder every day about whether to call the council to remove the litter or whether it belongs to the river and its litter people. Between 25 September and 30 September when I finally call the Council, there are infinite daily variations in the litter distribution, with the colorbonds disappearing, found located in a secret enclave at the back of the site, re-placed, litter removed, other litter appearing and reorganised into the ever changing installation. The more I play, the more embedded with litter I become, a litter-created body without organs.

I have now called the Council to come and pick it up. I struggle with this, it kind of belongs here and to the river. Sense of mourning its loss surprises. There is no communication through my invitation to respond, only the litter speaks. The Council eventually comes and one last visit to my installation reveals an empty site except for the two huge pieces of timber set up as seats. Grateful. I ban myself from the site for days, and return to my email for a letter from Feral Susan in response to the litter installation.

Emu Girrl

Back at litter site
school hols over
absurd relief at return of litter's people
reproliferation of river's litter
remnant boards from installation
not quite balanced on their large river rocks
a seat to contemplate flow of
river to its mouth at long ago
island home.

Email 7

Dearest Emu Girrl,

I have time (finally) to sit with your words and photographs. I adore the installations. What has happened to them? Did people respond to your call or did the installations disperse once again through unknown entanglements? My river is more secretive and only shows her secrets, her collections of things, after a roaring flood. After a recent flood, I picked up an empty glass Listerine mouthwash bottle with a plastic lid. I had never seen such a bottle. I searched for when Listerine stopped using glass bottles and plastic lids and learned that the bottle could have been anywhere from the 1950's to 1980's. I held the bottle in my hand feeling the weight of it realizing that it could very well be older than I am. And I wondered what the river left on it and what it left with the river. How did it end up in the river? Where did it come from? Did it travel from St. Louis, where Listerine was invented? Or further north? What did it see and experience bobbing along with shifting currents? What are the residues of its entanglements, residues I will never know? If the floodwaters had not washed it ashore, where would it have gone? And, then, I remember our conversations about Henderson Island, the repository of loads of plastic rubbish from every country in the world. You found the article compelling. The litter reminded you of messages in a bottle. Messages in litter stopped at a tiny island. All these messages carrying with them animals that have become entangled with them. The rivers create their own time and the litter materializes that sense of time that is a pure present – flows toward more flows, currents upon currents, carrying with them litter from todays and yesterdays.

One Fall day I found ripped out pages of a bible with highlighted passages around my house. It was a rather windy day and I watched as the pages danced across the browning grass and fallen leaves. I collected them, out of all the litter I have seen around my house, I've never seen ripped up bible pages. I picked up each of the bible pages and studied them. The windblown tissue-like pages felt soft under my fingertips even though the wind and elements had made the pages rougher, as if the previous entanglements had weathered them. How much of trash is reminding us of our momentary control of it? We place it in a bin. Make an installation. But, then, the trash moves … perhaps trash decenters us … a reminder of what we were, centered humans. On occasion, I don't clean the litter that sometimes makes it on my lawn through movements of wind. Each day I follow the trash, a plastic wrapper, a discarded fast food wrapper … and see where it goes and, possibly, to remind myself of how little control I have of it ... it is just a momentary entanglement with a gust of wind making it move elsewhere.

Feral Susan

Artwork tractor tyre
red and yellow stripes evenly placed around
Aboriginal colours, large black tyre,
but also one large arrow red outlined in green
pointing where?
Who knows, that is
the enticement of reading litter.

Email 8

Dearest FS,

I love your story of the wind blowing the trash into your yard, its movements like the flow of water that takes up the trash and carries it to different places. Last night there was a storm, one of our oogy boogy storms, middle of the night claps of thunder, flashes of lightning, drumming of rain on roof. Walking this morning air clean and cool, washed by rain, I come across the riverlands' own perfect trash installation. It is a beautiful, complex multi-textured multi-coloured intermingling of natural and artificial creations. All is held within intertwined entanglements of twigs, branches, leaves and logs arranged against the concrete span of the drainway. Colours are soft beige browns with flecks and patches of delicate faded red and orange where bark remains on peeling logs. Placed amongst these are the textures and colours of pale grey plastic bags, scattered shapes of worn white polystyrene, fragments of vivid orange plastic, faded blue of cardboard carton. It is as if everything is made more gentle by its slow becoming with water's flow. There is no loss, only the never-ending artful proliferation of the river's litter.

EG

The artful proliferation of the river's litter

It is the artful proliferation of the river's litter that calls Susan and Margaret to respond to the Mississippi River's multimodal creations of stories, music, song, art, and literature in all of their different manifestations. Litter flows from the great rivers of the world onto the remote Henderson island with the highest density of plastic rubbish anywhere in the world, mostly everyday things like cigarette lighters, plastic razors, toothbrushes, plastic scoops, and babies' dummies. It is here that our new litter-making identities come into being as we become Feral Susan and Emu Girrl, exchanging letters across the Pacific Ocean when we return to our own rivers to become with their litter.

Emu Girrl creates an installation with a disintegrating bike, plastic bags and bottles, glass bottles, aluminium drink cans, orange spray paint can, and other scraps of everyday life, underpants, metal frame of chair, metal rods, and perfect pair of pink-handled scissors. Feral Susan makes stories with a Listerine bottle picked up after a recent flood because her river is more secretive, only showing her secrets, her collections of things, after a roaring flood. She had never seen such a bottle, held it in her hand, feeling the weight of it, realizing that it could very well be older than her. She wondered what the river left on it and what it left with the river, how did it end up in the river, where it came from, what it saw and experienced bobbing along with shifting currents. And the nomadic orange chair becomes inseparably intertwined with the song of the lyre bird when it becomes the location for deep listening to the endless mimicking of the lyre bird's song in the cacophony of bird song in the river's lands.

Email 9

Dear Feral Susan,

Always in the middle that the writing/ideas happen. Imperceptible authorship. This quote from Barad (2012) seems to be perfect for our happenings and to describe the usness of our writing together.

> *Thought experiments are material matters. Thinking has never been a disembodied or uniquely human activity. Stepping into the void, opening to possibilities, straying, going out of bounds, off the beaten path—diverging and touching down again, swerving and returning, not as consecutive moves but as experiments in in/determinacy. Spinning off in any old direction is neither theorizing nor viable; it loses the thread, the touch of entangled beings (be)coming together-apart. All life forms (including inanimate forms of liveliness) do theory. The idea is to do collaborative research, to be in touch, in ways that enable response-ability*
>
> (Barad, 2012, 208)

Emu Girrl

Email 10

Dear Emu Girrl,

Barad's (2012) words are too perfect for us. Affective oogy boogy renders us imperceptible to ourselves in our thought experiments. Oogy boogy materializing from submarine communications cables resting deep in the ocean floor that help create experimental currents of thought. Vibrating oogy boogy moving through ocean currents delicately creating and recreating Henderson Island, our symbolic meeting place. Movements of the litter, bobbing with currents, materializes oogy boogy. Connective currents render us perceptible, then imperceptible, perceptible, and imperceptible again.

Consider all that divides humans, between ourselves and nature. Water renders those divisions into connections. Not only in terms of survival, but also in terms of a constant reminder that we are never so above nature. Water decenters us, pushes us in ebbs and flows until we reckon with the fact that we are only living with this planet for such a short time. We are creatures living and dying with so many other beings in the Anthropocene (Haraway, 2016). Water rips the crowns from our heads and pulls our bodies off our supposed thrones. When we're all in this together, my word, no one species rules. We are all litter. In touch. Doing theory.

Feral Susan

Note

1 University of Memphis, Tennessee, and the Mississippi River.

References

AERA(American Educational Research Association) (2018). Meeting of the waters: death and the global materialities of litter. Available at: www.aera.net/Events-Meetings/Annual-Meeting/2018-Annual-Meeting-General-Information

Barad, K. (2007). *Meeting the Universe Halfway: Quantum Physics and the Entanglement of Matter and Meaning*. Durham, NC: Duke University Press.

Barad, K. (2012). On touching: the inhuman that therefore I am. *Differences: A Journal of Feminist Cultural Studies*, 23(3), 206–223.

Bogue, R. (2009). Preface to multiple literacies theory. In D. Masny and D.R. Cole (eds), *Multiple Literacies Theory: A Deleuzian Perspective*. Rotterdam, the Netherlands: Sense Publishers, pp. vii–ix.

Colebrook, C. (ed.) (2010). *Extinction: Framing the End of the Species*. London: Open Humanities Press.

Cooper, D. (2017). Remote South Pacific island has highest levels of plastic rubbish in the world. *ABC News*, 16 May 2017.

Deleuze, G. (2006). *Two Regimes of Madness*, ed. D. Lapoujade, trans. A. Hodges and M. Taormina. New York: Semiotext(e).

Grosz, E. (2005). *Time Travels: Feminism, Nature, Power*. Durham, NC: Duke University Press.

Grosz, E. (2008). *Chaos, Territory, Art*. New York: Columbia University Press.

Haraway, D. (2008). *When Species Meet*. Minneapolis, MN: University of Minnesota Press.

Haraway, D. (2016). *Staying with the Trouble*. Durham, NC: Duke University Press.

Lather, P. and St. Pierre, E.A. (2013). Post-qualitative research. *International Journal of Qualitative Studies in Education*, 26, 629–633.

9 Regeneration

Of trees, weeds and tender intimacies

Introduction

One billion trees

In May 2019, the New Zealand government announced their goal of planting one billion trees by 2028 because:

> Trees promise to offer regenerative possibilities for landowners, communities, catchments, regions, and New Zealand, including helping to: improve land productivity; tackle environmental issues like erosion; reduce the effects of climate change by absorbing carbon dioxide; improve water quality; provide important habitats for a range of native species; enhance natural landscapes; provide another source of income from timber, honey, and carbon credits; create jobs and careers for our people.
>
> (TeUruRākau, Forestry, New Zealand, 2019)

In the context of anthropogenic changes to the Earth's climate, the planting and care of trees and other plants offer important regenerative possibilities for the planet. This, however, can only happen by the accumulation of millions of minor actions in very particular local places.

The chapter explores these possibilities in the very local site of bushcare regeneration of the riparian zone of Emu Green by the Bass Fishers of Sydney. The fishing club was allocated their own regeneration spot on the Nepean River at Emu Green in return for access to the waterways to engage in fishing and release of the native bass fish. Because of their far-reaching involvement in rivers and waterways, the Bass Fishers are unique in their approach to bush regeneration. For them, it is not only about the trees and weeds, and all the plants in between, but about the relationship between the waterways and the riparian zone. I joined the regeneration work of the group seven years ago when I came across them on my daily walk. The chapter explores the multiple entangled relations between the plants, the humans who labour there, the waterways, and its creatures, through long conversations recorded with the Bass Fishers, and my own learning experiences in bushcare riparian zone work, through the advice given by the Bushcare official (Gasparre, 2012). The iconic Rhine River flows into and through this chapter in its question about how our rivers can be transformed through the care of their riparian zones.

Bushcare and bush regeneration

There are many alternative names for the actions of volunteers everywhere in caring for their local wild places, the key terms that conceptualize this work are 'bush', 'care' and 'regeneration'. In the entangled relations between plants and humans in the riparian zone, the questions are: who or what is being cared for, and who or what is being regenerated? Are plants caring for the human and everything else that dwells there, or are humans caring for plants and everything else that dwells there? Or are they all becoming together, in the Deleuzean sense?

Bushcare

The bush has a special symbolic meaning in Australian life, often meaning away from the city, but when used to describe the land, the bush means a wooded area, but not dense forest. It is usually dry and nitrogen-poor soil, mostly grassless, with thin to thick woody shrubs and bushes, with some eucalypt trees. Bushcare is a national movement which started in the suburbs of Sydney in the 1980s. This environmental volunteer work (sometimes called Urban Landcare) occurs mostly on public land: in bushland, wetlands and around waterways managed by local councils or NSW government agencies. Environmental volunteering programmes such as Bushcare have steadily grown since their inception and are now an accepted and successful partnership between many government land management organisations and community members.

Bush regeneration

Regeneration has many meanings that are evocative when applied to the activities of bush regeneration: Regeneration involves remaking something; *Regeneration* means 'create again', humans experience regeneration when they produce new skin cells to heal wounds and burns; Regeneration is also a religious term that means a 'new birth' in which a person changes from sinner to true believer; forming again (especially with improvements or removal of defects). Regeneration can mean renewing and reconstituting; resurgence, revitalisation, revival, revivification; growth anew of lost tissue or destroyed parts or organs.

Riparian zone bushcare regeneration

Riparian zones are very special places for regeneration activities at the interface between aquatic and terrestrial environments. Healthy riparian zones are critical to aquatic-terrestrial ecosystem function and, ultimately, to waterway health. Riparian areas are vulnerable and easily degraded because their physical connectivity leaves little to buffer them against impacts occurring in adjacent ecosystems. Expanding urban development increases pressure on catchments, aquatic and wetland habitats, and the remnants of native vegetation, and fragmentation of the riparian zone habitat reduces the value of these zones as corridors connecting different areas and habitats.

Regeneration

Regeneration site: winter/spring, 2018
Lush green plants become
bundles of dry twigs poking
out from protective guards
gone dead in a long cold dry
winter of now rain.
Let's cut them off I say,
pull them out get ready
for new planting in Spring.
Just keep watering them, they say
over and over again, *just keep watering them*
until finally the spring rains come
and the dry twigs sprout
new bright green leaves
tender tended loved.

Every time I visited the plantings in that long cold dry winter, I mourned the loss of the many small plants as they turned into bunches of dry twigs poking out of their tree guards. They were dead. Let's pull them out, I'd say, and every time the same answer, just keep watering them, just keep watering them, they'd tell me. Then the spring rains came and the dry twigs sprouted new bright green leaves, some from their base, some from their stems and some from the top where they started anew. I talked to the plants, *I thought you were dead, what can you teach me about regeneration?* I learn their names, both common and Latin names, so I could address them properly: Cheese tree: *Glochidion ferdinandi*; Australian indigo: *Indigophera Australis*; Coffee bush; *Breynia oblongifolia*; Forest red gum: *Eucalyptus tereticornis*, and I promise to find out how this miracle of regeneration happens.

I learn that this regeneration is unique and comes from the time of Gondwanaland when Australia was still connected to Antarctica. All Australian plants have adapted to a fire-prone environment, along with regular droughts, and nutrient-poor soils, and most can re-shoot from protected buds on their stems or roots, so they can recover rapidly after a fire. They have the ability to regenerate after severe loss of biomass during drought, when they are eaten by native animals or attacked by insects, as well as when burnt by fire. They re-sprout from special cells called meristematic tissue that can make new growth from apical, epicormic, basal, or belowground buds. These ancient adaptations helped them to adjust to vast changes in climate over millions of years and continue to bring them back to life in their place by the river. Through such conversations and countless acts of bushcare at monthly working bees, I learn the story of bushcare, and the Bass Fishers' unique approach to riparian zone regeneration.

Bushcare: the bush's love story

In the 1960s, the 'eccentric' Bradley sisters in Mosman started working to restore their 'local patch' of bush, developing a series of weed control and native vegetation recovery techniques through trial and error. The 'Bradley method' involved methodically clearing small areas in and around healthy native vegetation so that each area was re-colonised by the regeneration of native plants, replacing the weeds. Their work was the beginning of bush regeneration in Australia.

Bass and the bush

FISHER 1: When you're fishing for bass in the summertime and you're using surface lures, you always generally fish under the overhanging branches of trees. Yes, that's because during the day, in particular, that's where the shade is, and the bass don't like to be out in the sunny water, they always head for the shade, so that's why we do it, and I think it definitely does improve the habitat. With mass planting, you know, lots of shrubs and trees close together, again on the bank, you know.

FISHER 2: I went to Fishers for fish habitat rehabilitation, did a two-day study day, really enjoyed it. This is why looking after the environment's better than stalking, because if you just stalk. the fish die and you don't get any recruitment of new fish or juvenile fish, but by enhancing the environment, the fish have a place to live and they can recruit themselves. So, the interest in bush regeneration was to give something back to fishing, to make the fishing more productive – from the years of degradation that's happened through bad management of rivers. So, it all begins to tie in on itself. It's not just one thing leading another, it's like a perpetual circle. It just keeps feeding.

BUSHCARE OFFICIAL: A Bushcare site is a living, adapting and constantly changing thing. Bushcare and other on-ground work are part of the big picture of biodiversity restoration across Australia. The programme encourages recognition of the interconnectedness of work across all sites, big or small and its contribution to habitat health and landscape change. Today, over 17,000 volunteers are working across Sydney to protect and restore bushland, waterways and other natural areas. Bushcare, Landcare, Coastcare, Reefcare, Streetcare, Parkcare, are about groups of people looking after natural areas over a period of time.

The perpetual circle

FISHER 1: Well, for the fish, it attracts – it's not just fish either, by the way, it's also for the birds – but it attracts insects. And 40 per cent of bass feed comes from insects falling into the water. So, with more native trees around, it attracts more insects, more birds, the birds attack the insects, they fall in the water, it's food for fish, so it's a win-win for us. It's a whole little cycle, eh? It's not just for the bass, if we have more trees planted along the banks, it also helps the habitat for platypus. I haven't seen them here, but I have heard that they've actually been up around Penrith weir, so they must be in the river here somewhere.

FISHER 2: I think it does, in the little pocket that we're working on. But I think it's part of a much bigger picture. Well, the bigger picture really is that we're a bunch of keen fishermen who are more concerned about the environment and what it does, that's the major reason, because there's so many noxious weeds; balloon vine, lantana, stuff like that, choking the life out of our riparian growth that we need to do something, because it enhances our native fish, which is one of our ultimate goals.

FISHER 4: The trees that we've planted ourselves, which were four years ago, some of them are now 5, 10 metres tall. I didn't know they grew that fast and they're now forming their own canopy. They're now providing shade underneath, hopefully for native animals, native insects, native plants, and their presence is helping to keep away invasive species, as I understand.

FISHER 5: I guess, if you plant – especially along the waterways – the seeds and stuff will travel along the waterways and it's not only here but the seeds will travel and may germinate somewhere, 100 metres, 100 kilometres down the river, for instance, that they can, and it'll give the biodiversity all along this river and not only to this spot that we're doing.

BUSHCARE OFFICIAL: Knowing what animals live at your site, or that might make use of your site occasionally, is a very critical step in getting Bushcare right. Think carefully about the needs of wildlife on your site, not just the plants! Wildlife sometimes don't discriminate between native plants and weeds. So clearing out all of the weeds in one go may not be the best approach. Wildlife also will not like 'tidy' areas without dead branches or leaf litter. These are important for protection, shelter and food sources. Avoid the temptation to turn the bush into a park-like garden. Diversity of plants, vegetation structure and mess are all good. Bushcare efforts need to always keep in mind the needs of the local native animals.

Step by step: as slow as a tree grows

FISHER 1: Yeah, step by step, and it's taken us two hundred years to get into the problem we have. We're not going to fix it overnight, but you know, it's like every journey, it starts with the first step.

FISHER 2: A net gain is probably unlikely because it's quite a small area considering the length of this river, but the fact is that if we can do it and it shows to be working and people can see the benefit, then hopefully that thought pattern or that principle will carry on and multiply and more people will be more concerned.

FISHER 3: I don't know how much guarantee you can put on the bigger picture but certainly doing nothing is – there's no bigger picture or a long-term future, but by doing stuff you are creating a bigger picture for your future and if you move away, you take that experience with you and you've got a bit of a voice with some experience to speak up about and maybe get other people involved.

FISHER 4: I can now appreciate other people's projects in rehabilitation because of the volume of work that it takes. You don't just go down for 2 hours and it's done. It's month after month after month and there is a boredom factor creeps in, so in order to keep working, if you're working with other people who are interested in fishing, are they interested in your family?, are they interested in your job?, and you're talking about other stuff, then it maintains your focus.

BUSHCARE OFFICIAL: In the spectrum of approaches to improving our natural environment we often talk about 'The 3 Rs'. It is very important to understand that restoration of a natural area is a long-term process. *Retain*: Conserve existing natural areas as a first priority. *Regenerate*: Assist the bush to restore itself by not clearing weeds too quickly, allowing time to wait and see what plants may appear from the stored seed in the soil or be introduced by native fauna, especially birds. Weeds are removed at appropriate times, e.g. before setting seed. *Replant*: Planting should only be considered after a site's natural ability to recover has been assessed as poor, such as where the natural soils have been removed or extensively disturbed. Only plants grown from locally collected seed are used in revegetation programmes.

Planting and learning the language of plants

FISHER 1: I've learnt a little bit about some of the species that we've planted, and times to plant, and times not to plant, and I think it's been good from that point of view, and, I mean, I have a fair bit of experience with planting natives from home, you know, but I've learnt a few new things here that I didn't know, like how deep to plant them, and you plant them 50 mm below the surface, so there's a natural pool. If there's rain, an actual pool is catching it, and little things like that.

FISHER 2: Yeah, I love it. I love coming here and love to see the plants grow. There's some trees that we planted first here that are about 8 to 10 metres tall now, so it's good to see what we've done, the work that we've done, that what we do works, but still there's a lot of maintenance but that's something that we have to keep up for a while. Hopefully nature will take over one day and keep the weeds out and, yeah, let the native plants grow.

FISHER 3: It helps to get all the weeds out, the non-native plants basically. We try to grow native plants which are from the area, from seeds from the area, and we try to grow them here on site and that'll help eventually all the wildlife and bring back, hopefully, a bit back – the nature back – like what it was, what it should have been.

FISHER 4: I'm originally Dutch so I don't know all these plants, don't know half the plant names from the plants that we planted, yeah, and it's part of my work anyway, so, it benefits me at work as well too, to determine what's a weed and what's not a weed.

FISHER 5: I do not know the names of some of our native species, in fact, I'm probably more familiar with the names of invasive species, and I've learned how to identify some of them. So, from a plant point of view, I know a little bit more and I think there's lots more to learn and I'm kind of interested to learn that I just need to start that happening – reading more books, and I have bought books from CSIRO on plants with the view that I'll start reading them, so that's something I'm planning to do.

BUSHCARE OFFICIAL: Walk through the site with someone who knows plants. The common name is the name that many people apply to a plant that does not use scientific terminology, e.g. Sydney blue gum, grey gum, stringy bark, etc. A botanically correct name is applied to a plant that will be recognised as the only correct name for a particular species of plant. This includes a Genus (always with the first letter a capital) and a species, e. g., *Grevillea juniperina*. A grouping of plants that grow in similar site conditions and have similar characteristics that together can be recognised. It is based on the structure of the plant association and make-up of species on a given site, e.g., saltmarsh, Sydney turpentine, Ironbark forest.

Weeding transformations

FISHER 4: I can still see a fantastic change. I now know about some invasive species like the balloon vine, and when balloon vine covers trees, it blocks out the light and those lower branches get denied light, they don't grow any leaves, and they pretty much look dead until you look at the top of the tree and you actually see there's green leaf and the tree isn't dead and then it blankets out all that light, you don't get birds, you don't get so much insects, it's almost nearly eerily quiet and dead and it's just this blanket of balloon vine and when I saw that, we'd come in and we had, I think they call it curtain cutting or something, when you just cut it all at head height and then we poison the roots and all that foliage dies away and dies down and then falls down over time and then we could see more light coming through.

FISHER 1: Well, we've removed tons of weeds, including lantana, privet, balloon vine, and replanted with native trees. And I suppose you really have to see it to believe the difference it has made to this site.

FISHER 2: I think what happens here is the environment was overrun by noxious weeds and trees, and we've managed to change that here and remove a lot of that, and introduce natives, which now make the area look a little bit more native than it was.

FISHER 3: The bigger picture, really is that we're a bunch of keen fishermen who are more concerned about the environment and what it does, because there's so many noxious weeds; balloon vine, lantana, stuff like that choking the life out of our riparian growth that we need to do something.

BUSHCARE OFFICIAL: Many volunteers are now donating large amounts of their time to weeding and restoring bushland remnants and it is a laborious business. In many remnants, weeds such as lantana have replaced the native understorey. These weeds are now important habitat for small birds as they now offer the only dense shelter sites. Removing these weeds quickly therefore destroys the habitat for small birds, which is one of the reasons we value remnant bushland. Best-practice bushcare requires the gradual removal of weeds, working from the areas of lowest to highest weed density. This means that native plants have the chance to re-establish a dense understorey before the thickest patches of weeds are removed.

Building community and tender intimacies

FISHER 1: You can now see those little husks that I was telling you about from the cicadas. You can now see lizards. There's little colourful tree birds fly through. It's not just your big crows or magpies. You're getting these little ones and, in fact, the last time I was here, I commented on a couple of things that I saw that anybody could see which was – at the water's edge there was kingfishers on my left, flying from a branch down into the water and pinging at fish and flying back up. Another thing I saw was a huge monitor lizard that had been drinking at the water and as I walked down, it took off and climbed up a tree. You see more snakes because the place is more open. You can feel the breeze. You can feel the wind.

FISHER 2: I commune with nature enough in my pastime, but this way, I get a group, get together with a group of my club mates, we work, we have a BBQ, we talk non-stop, and it's not a bad day. Sometimes we bring our, you know, some of our younger members might bring their kids out. I've even brought my grandson out, so, you know, yes, I quite enjoy it.

FISHER 3: It means that I have friends now here in Australia. It was hard to make friends in the beginning but now being involved with a club and – especially this club, it's like almost kind of a family now, so it's always the same group so you get to know each other really well and, yeah, it helps me to be involved with something that I love and not just somebody that comes from the Netherlands and just sits there. I love being involved with everything so that's why I guess I'm president now too. I don't mind it. They talked me into it but it's good. It's good.

FISHER 4: My daughter who's two and a half has been coming here since she was about – just before one, actually quite likes it and without me prompting her, she picked up my fishing rod and started winding in the reel. She likes lying about in the grass and near the water and she's seen herring, mullet, and bass all caught from this area and if all the native species were gone, she might only ever see carp. It's actually a nice place to bring the family, there's now light, soft grasses, the sound of birds. You can actually hear the water. I can bring the family and they can muck in and pull some weeds or do some watering and my daughter likes to do the watering, I hold it and she just pours it. We can bring family, you can bring friends, you can bring your pets as long as your dogs don't run wild.

FISHER 5: I love this. I love nature and I love to be outdoors. You don't hear any city noise. We don't hear a thing, just a bit of the water at the background and the birds and the crickets and the cicadas and stuff like that. So, yeah, it's good to be out and be here.

FISHER 6: We could see smaller birds from forests permeating and we could hear more birds and we could actually see through the trees through to the river and then, as a result of that, those lower branches had more leaves on them and you could see the shells of cicadas whenever they come out and crawl up the tree and then they moult and fly off. So, we could actually see that there was more insect life. You could see lizards on the trees that weren't there before, sunning themselves. We see more butterflies coming through.

FISHER 7: If you've got healthy trees, healthy grasses, that are native, they will support native insects and native insects are part of the bass diet. There'll be more biomass because we've now got native plants, getting exposed to sunshine, having native foliage, support native insects, you've also got additional things like native birds. So, native birds can keep down pest insects possibly, but if there's native insects for fish, there'll be more native insects for birds.

FISHER 8: On Sunday, I went for my 5th morning session over June/July and, together with some earlier more random observations, I've now made it to 50 native species of birds (plus 4 introduced!). Also couldn't resist sharing what I reckon are my favourite pictures to date. A pair of pardalotes that I reckon must be building a nest under a stack of pulled lantana.

Community for Bass Fishers is made up of trees, weeds, plants, fish, flowers, light, birds, insects, butterflies, lizards, snakes, children, grandchildren, pets, the sound of water, cicadas, birds, crickets, the feel of the breeze and the wind, and the conviviality of human friendship joining together in slow time, the time for the planet through which these tender intimacies are created.

Making community

Casuarina tree

Tired bones resting against nesting tree
quiet bodymind at rest

tinkle bell birds squark of distant cockies
twitter of little birds.
Where's humans gone
better go back to planting
casuarinas, prickle bushes for little birds
ti trees wattles
in sandy hydrophobic soil
with water well around hole
3 finger pinch water crystals and bed baby tree
in powder dry soil tamp firm
dip bucket into river with
rustle little fish to water.
Bone tired now 500 trees planted
sausages yummy chatting too.

Resting, nesting, casuarina tree

Sausages, onion, salad, bread and tomato sauce soooo welcome as we sit in the shade of the big casuarina tree, because hard work makes us hungry. One of the fishers points out a large yellow-banded wasp at its mud nest on the trunk above my head. Another photographs a cuckoo bird. Then the leader sits down next to me at the base of the tree, a seat formed by root mass and dry needles. He talks about Xylem Water Solutions and how they want to water the site on 22 March. We talk about the fish records. The river is divided into five river regions for the purpose of data collection, starting from the Hawkesbury around Windsor and going all the way to Camden in south-western Sydney. They record everything they catch in two fish catches a year. Their records of native fish in the river system have been now collected for 27 years, the longest-standing record of native fish in Australia.

It is the gathering around the casuarina tree for morning tea, lunches, relaxing and chatting that keeps us all going, but much learning happens there too. Today it's a planting day and Al Fowkes names all the species waiting in their tube stock pots and how they should be planted. There are wattles, indigoferas and grasses that need to be in small cleared spots between the bushes and trees, where their seeding will help them to spread. The tiny plants wait in the shade of the casuarina as we come and go taking small bundles of plants to place into their new riverlands location.

Planting and seeding

The planting is hard, hot, back-breaking work. Getting up and down and up and down all the time, to dig holes, plant acacias and indigoferas, form wells, position and place tree guards and stakes in the river's rocky ground, it is all of the river. I plant four acacias in the riverbank. Digging in dry stony ground, digging out stones and some rubbish, bricks and other semi-decomposed plastic items. Each hole is ready for planting and a tube stock acacia placed beside each one. Water crystals are placed in the bottom of the hole and stirred into the dry ground, the tube stock is removed from its pot and the bottom of the roots is lightly tickled out, with their hole made larger by the removal of big and small river stones. Soil is replaced around each tiny tree and tamped in by treading with feet around it to create a well so water will soak into tree roots rather than flow away. The river stones are placed in a circle around the new trees to hold the water in their water well.

Seeding: the kangaroo grass project

> *Al Fowkes, 14/11/18* Hi, Margaret, Was just wandering through my local reserve and noticed that the Themeda (Kangaroo Grass) is carrying heaps of seed. I remember last year we missed the boat at our site when that large patch to the north of the airfield was mowed. Do you reckon you could check it out next time you're down there? If there's heaps of seed, we can collect some next session or, if you're keen, you could do it before then. Simply strip off seed with your hand (if it's ripe enough, it will come) and place in a paper bag. Seed collecting etiquette is not to take more than 10 per cent from any one plant. Put the paper bag somewhere hot for a week or two (like the dashboard of your car) and it will be ready to spread around our planted areas. Would be a worthwhile experiment and a good source of free plants if it works. What do you think?
>
> Alan

> *Margaret, 15/11/18* Hi again, Al, I have checked out the kangaroo grass and attach two images. They are not very clear but you may be able to get the gist. My impression is that the heads are a mix of too brown and old from last season, and too new and green from this season, can you advise whether either or both will produce viable seed? Thanks.
>
> Margaret

> *Al Fowkes, 15/11/18* Hi, Margaret, Thanks for the photos and I think your assessment is spot on. Having said that, if you're walking by, it won't hurt to bag a few of the old seed heads and we may get a pleasant surprise. Hopefully the new ones will be ready in time for our December session and we can collect a substantial number then. Keep well.
>
> Al

And so starts the kangaroo grass project. I have long loved kangaroo grass for its beautiful change of colour from bright green to red and to dry light brown as it moves through its seasons of growth. I have photographed it as it sways in the breeze, a perfection of soft movement in tune with the movement of air and weather. But I know nothing of its languages, how to collect the seed or how to prepare it for planting. I learn that kangaroo grass is *Themeda Triandra*, previously *Themeda Australis* until it was discovered to be genetically similar to grasses in Africa and Asia as well as Australia. Aboriginal people of the tablelands collected them in great quantities, to grind and bake into cakes. Kangaroo grass was gathered in wooden bowls, the seeds were separated by winnowing, and grinding stones were used to produce flour which was mixed with water and cooked to make damper.

I come to know the seeds from my intimate engagement with their form as I try to determine when the new heads will become ripe enough to collect. The seeds are arranged along a fine tough stalk and their weight makes the stalks bend over at the top and sway in the breeze. Each seed head has a number of seeds encased in a protective cover, a 'caryopsis' and each ends in a long fine spike called an awn. The caryopsis changes from bright green when new, to rusty red as it matures, and dry brown as it dies. The awns are black but similarly change colour until they too wither and die. Their mass appearance is beautiful as they change with the season. To collect them, I need to know when the seed is mature enough to be released from its caryopsis but not too old. I collect bundles of heads and arrange them in a pile on an outdoor table to observe their form. Each day another bundle is added until I can see that they are at the right stage of maturity.

I place the heads in paper bags and carry them in a shopping bag to the seat of my car on a hot summer's day. When I arrive at work, I park carefully so that the dashboard is exposed to maximum sunlight. Then I arrange the paper bags in a row along the dashboard as Al Folkes has advised. In that instant an extraordinary encounter emerges. The bags start popping. I hop out of the driver's seat to get my iPhone from the back to record the sounds and they pop even more, pop pop pop they go in their paper bags. They speak to me, telling me of their readiness to be planted.

The next day, I take them down to our bushcare site and sprinkle the seeds with their awns still attached, all around the grassy area near our nesting resting tree. They will take 6 to 12 months after sowing to germinate and then the tall grasses will appear long before they grow their beautiful characteristic kangaroo grass seed heads. I have not seen it growing in our patch yet but its long swaying heads self-seed in my native garden, leaning over the road where a local neighbour collects it to feed ailing joeys, the babies of the kangaroo that gave it its name.

Watering

Its midsummer and the plants need water, so I walk to the river with two buckets to collect water from the edge of the river, where I hold onto Damian's fishing stick stuck firmly into the ground with a prong at the top so as not to drop into the water from the grassy overgrown bank. Lean over dip and fill, but when I look at the water in the buckets, it is full of tiny black fish. I ponder whether to just water with them, thinking there are millions of them in the river and walk to the plants, but when I get there, I just cannot do it. Cannot tip these vital living water creatures onto dry ground to die. I take a video of their black wibbling as they swim around in the bucket. Are they mosquito fish, or gudgeons, the little native fish, and what sort of gudgeons? One of the Bass Fishers says he noticed lots of gudgeons in the channel. They are a sign, he says, that the bass will be there because they feed on a particular type of gudgeon. He described rainbow gudgeons whose fins go bright red at breeding time. So I trudge back down to the river, tip the tiny fish back into the edge, then wade across the stones to where the current flows and no tiny fish are there. Dip and fill, dip and fill in the river and back to the trees. It's the slow time of watering trees with a watering can from the river. The slow time of trees growing, slow time of the planet.

Tender intimacies

The world of bush regeneration is full of critters of every different sort and size, land dwelling, water dwelling, ground dwelling, air dwelling and all the spaces in between. A tiny spider hurries across the ground of the plastic tent that protects our newly planted trees. Each tent provides a mini eco-system and I notice, even when I have not watered, that droplets of moisture have condensed on their sides and run back down to the parched soil at the base. A butterfly lands on my hand as I kneel on the ground, bright blue wings on brown, and flutters to land on soft casuarina-needled ground.

A heron stands at the pool where I come with my buckets to collect water. It walks daintily on delicate fine legs in the shallow water dipping its head just as I dip my bucket, dip and lift, dip and lift, until it flies off past the little island. The white egret is here too, just sitting where the white egret sits. The life here today is ducks, grebes, a large silver fish jumps out of the water. Cormorants, birds, and even the flies seem part of the rich proliferation of life here today.

Damien is here with his small daughter Gracie, who comes with him to our regen days. Flies are bothering her face, he wipes them away and only more come. Schools of small fish riffle the still surface of the water as Damien feeds bread to attract them. He says they are mullet or herring and explains how the fast flow of a flooding river enables small fish to come into the little inlets and crevices to feed on worms and other land creatures. He catches three small mullet, each one bigger than the last, yellow-eyed mullet, and tries to entice Gracie to come closer. She is cautious and when one wriggles out of his hand and lands on her foot, she cries, only briefly and soon will touch the smooth cold scales of the fish with my hand holding hers. So soft and small.

Destruction and regeneration

There are so many stories of 'the hoons' as the Bass Fishers call them, in this spot where they love to drive their four-wheel drives, often leaving a trail of damaged trees. They have driven over mature trees, smashed down smaller bushes, cut down a large casuarina by the river with a chainsaw, inscribing their insignia on the remnants of the trunk. My whole body responds viscerally to this violence, as if I am part of these violated trees and bushes, and yet, over and over again I am renewed by the ability of these plants to repair. A 12-metre-high eucalypt is rammed to the ground, driven over, torn trunk flattened, whole tree and leafy branches lying along the ground. As slow as a tree grows, demolished in a few minutes of vicarious violence.

The flattened tree dies slowly over the next weeks, its tough greeny-grey leaves turning a dry crackly, lifeless brown. Then slowly as a tree grows, the eucalypt starts to re-sprout along its dying stem. First, tiny, soft, tender new leaves appear, bright red growing tips as if it is the season of growth, then the growing tips slowly grow into small branches of their own. I revisit this tree over and over again, victorious in its regeneration, until finally, it is finished; the three new branches are now nearly as tall as their nearby intact friends. The dead trunk still lies along the ground, no longer firm but soft, worm-holed, home to a bright orange fungi growing along its decaying core, becoming part of the nourishing earth once more, a part of the perpetual cycle that fungi, lichen, insects and even the four-wheel drivers who love to destroy, create in this place of regeneration.

Prickly intimacies: weed's regenerative powers

Weeding is the most difficult and soul-destroying work of the riparian zone where the proliferation of weeds seems to far outweigh our capacities to remove them. Farmers friend, or *biden pilosa,* is one of the most discouraging of all. I'm buried in a mountain of farmers friend, long past their small green leafy stage, these ones are so tall they are higher than me, growing in great massive clumps in the plantings close by the river, dry brown and prickly. Tall and ugly, covered in black prickles, they bury smaller plants that need to be rescued. My Penrith City Council branded shirt is long gone, discarded after a session of biden weeding with farmers friends turning me into a prickly echidna. I now wear overalls but still they find places to insert themselves in my hair, falling into the neck, finding my socks, and covering my gloves. They are all over me again. Working alongside them, I slide on the ground through the massive weed infestation, trying to work out an approach that will not find me buried alongside them.

The routine is to remove the seed heads carefully so as not to spread the seed but I figure they have already showered themselves all over the place. So I pull, cut and pile, pull, cut and pile, sliding from clump to clump, getting dirtier and dirtier, hotter and hotter, but rescuing many small plants. There are ti trees buried in the mess, tiny prickle bushes that we planted for birds, cheese trees, small wattles, and even some eucalypts that have fallen prey to the dominating *biden pilosa.* Damien jokes about removing the last biden, then one day when I want to write back to his joke, I decide to check on the spelling. Such a simple thing, is this 'bidun', or 'biden', and *biden pilosa* takes me on the most extraordinary world tour. *Biden pilosa* has colonised every country and language of the world with its common names recorded in 18 languages, as well as having many names in English, *beggar's tick, beggar ticks, hairy beggar ticks, black jack, broom stick, broom stuff, cobbler's pegs, devil's needles, hairy bidens, Spanish needle, farmers friend, Devil's pitchfork.*

The healing power of biden pilosa

In another surprising turn, my intimate prickly friend has celebrated healing powers. In traditional Chinese medicine *biden pilosa* is used for clearing heat and removing toxins. It is extensively sourced among many traditional healers of Africa and Hawaii. A juice made from the leaves is used to dress wounds and ulcers; a decoction of leaves is anti-inflammatory and promotes good health, while the whole plant is anti-rheumatic and is also used in enemas to treat intestinal ailments. Substances isolated from the leaves are bactericidal and fungicidal and are used for the treatment of thrush and candida. Even *biden pilosa* compensates for its prickly intimacy with its gift of regenerative healing.

The hard work of weeding, watering, planting, seeding has its many rewards in the regeneration of the rivers' lands. The flowers that appear in the season of *Ngoonungi, Murrai'yunggory*, offer their gifts of great beauty, in their own eternal cycles of blossoming, pollination and renewal.

Flowers celebrate our bush care

The Dharawal season of *Ngoonungi, Murrai'yunggory*, of cool getting warmer is a time for ceremony when the clans gather to celebrate a time of plenty (Bodkin, 2017, 19). At this time of regeneration and renewal, new leaves spring forth in bright colours, the bright pink growing tips of the callistemon, the shades of pinky orange and dull red of the eucalypts, and light greeny yellows of many others, all their new leaves shining bright and tender into the new season. The wattles burst into flower, the long pale yellow flowers of blue-leafed *Acacia longifolia*, the fluffy light gold of *Acacia floribunda*, the brilliant golden balls of *Acacia decurrens*, the green wattle. The brilliant pink of indigofera, the pale gold of bottlebrush, the fluffy white of the clematis, the mauve and purple of the white cedar tree, and many more. I record and photograph them all for our celebration of renewal and learn about their seasonal regeneration.

The green wattle shines gold

The green wattle stands tall and spreading behind our nesting resting casuarina. In this season of cool getting warmer, it shines with a mass of fluffy golden balls spread throughout its fine bright leaves and dark brown twisted branches. It is *Acacia decurrens*, with multiple sprigs of deep gold, so different from the other quieter, more modest, wattles that have already blossomed. It is *Acacia decurrens* and its golden-yellow flowers are bisexual and fragrant. The golden colour of the flowers signifies their rich source of protein, designed to attract butterflies, beetles, wasps and native bees, who spread the pollen from flower to flower, cross-fertilising them. Flowering is followed by seed pods with dark brown or reddish brown to black seeds lying inside a parallel-sided, flattish, smooth pod until they are dispersed by ants or birds, forming a seedbank in the soil. The green wattle provides unripe seed for sulphur-crested cockatoos, and the foliage serves as food for the caterpillars of the double-spotted line blue butterfly, the moonlight jewel, the imperial hairstreak, the ictinus blue, the amethyst hairstreak, and silky hairstreak butterflies, all likely pollinators of these and other flowers.

Indigofera *opens to a sparkle of purple pink*

Indigofera, Australian Indigo, bursts into its spray of glorious bright pinky mauve, shining in the morning sun against its dark green pinnate leaves. Shyly hidden at the back of the plantings, it has the shape and structure of a miniature tree, branches splayed out from its centre, flowers sprinkled throughout. Up close, the flowers are small pea flowers, tiny pink love hearts with darker purple interior. More and more appear with their showers of vivid purple pink in dark rich green, from their hidden places at the back of the low bushes. They are *indigofera australis,* the Australian indigo or Austral indigo. *Indigofera* is neo-Latin for 'bearing Indigo', a purple dye originally obtained from some *Indigofera* species. The flowers are a source of pollen and nectar for many native insects, including bees and wasps and the plant is also a useful food plant for butterfly larvae. After flowering, the shrub is hung with shiny, bright-green pods ripening to become firm and red-brown at maturity with up to ten hard, grey-brown seeds adapted to be activated by fire, as with many Australian plants.

The white bottlebrush glows into late evening

The pale cream, fluffy flowers of the white bottlebrush are opening here and there with just a few flowers, but a small bush stands straight and tall against the background of a pink and blue sunset-reflected pool. The upside-down reflections of the trees on the opposite bank are mirrored dark on the water's still surface. In the centre foreground it is the white bottlebrush that owns the scene, massed fluffy, pale, greeny-gold brushes shining in the last of the evening light, it's the perfect expression of the season of cool becoming warm. It is *Melaleuca salicina,* commonly known as white bottlebrush or willow bottlebrush, *Callistemon salignus. Callistemon,* from Greek *kalos* meaning beautiful and *stemon* stamens, *salignus,* from Latin *salignus,* of the willow, referring to supposed similarity of the leaves to the willow, *Salix.* Each flower is made up of masses of fine creamy filaments with a tiny globe of pollen on its tip producing its creamy white colour that shines in the late evening. As the season moves on, each flower will produce woody fruits that form in clusters along the stem that are held on the plant for many years until they are stimulated by fire.

White clematis shines white in dark green tendrils

The white clematis is flowering at the base of three casuarinas, dark green leaves lit by the myriad of white stars. Another festoons the twisted branches of an old dead wattle with its white starburst. Close up, the stars have layers of white fluffy stamen emanating from their inner centre with the faintest tinge of gold, some still closed white buds. It is *Clematis aristata, Clematis* from the Greek, *klematis*, the name of a plant, and *aristata* from the Latin, meaning bearded, known by various common names including traveller's joy, goatsbeard, and old man's beard. The flowers are either male or female with either silky stamens or slender, plumed styles that persist on the female flowers to form curved feathery appendages on the tightly clustered fruit. Fertile male and female reproductive structures occur in flowers of separate plants making this species an obligate out-crosser with pollen movement among plants facilitated by insects and wind.

One last flower, the white cedar

It's the giant white cedar just coming into flower. Exquisite clusters of buds in delicate pale pink, purple and white with crystal-clear droplets of the gentle spring rain. Some flowers are newly opened, five palest of pale pink petals arranged around a maroon tubular holder for the nectar-filled stamen within, an open petalled invitation to their insect pollinators. A fine fragrance already emanates from these open flowers filling the air with their readiness for reproduction. It is the late season flowering of the white cedar, *Melia azedarach*, commonly known as Persian lilac, Chinaberry and umbrella tree. *Melia* is the Greek name for the ash tree, an allusion to the similarity of the leaves; *azedarach* is from the name of an ancient poisonous tree, *Azadaracht*. After flowering, the tree is covered with round golden seed pods that drop to the ground, producing countless tiny seedlings that spring up under their deep canopy. The white cedar is one of the many plants that self-colonises the riparian zone, becoming its own agent of bushcare regeneration.

Transforming our waterways

The renewal of plants in their new growth, blossoming, pollination, and seed production in the time of *Ngoonungi, Murrai'yunggory*, of cool getting warmer, celebrates the transformation of the riparian zone, a transformation parallel to that of the River Rhine. Terrestrial and aquatic life forms come together in this edge place as one of the richest places of biodiversity on the planet. The human is recreated in the tender intimacies of care for plants and all that lives, transforming lives, regenerating communities which are made up of trees, weeds, plants, fish, flowers, light, birds, insects, butterflies, lizards, snakes, children, grandchildren, pets, the sound of water, cicadas, birds, crickets, the feel of the breeze and the wind, and the slow time, the time for the planet through which these tender intimacies are created.

References

Bodkin, F. (2017). *D'harawal: Climate and Natural Resources*. Sussex Inlet, NSW: Envirobook, Trekaway Pty Ltd.

Gasparre, F. (2012). *Bushcare Boosters Module 2: Bushcare and the Big Picture*. Eastwood, NSW: Bushland Management Solutions. Available at: www.aabr.org.au/images/stories/resources/manuals/VCN2012/BBM2_ParticipantsWorkbook.pdf

TeUruRākau, Forestry, New Zealand (2019). About the One Billion Trees Programme: Why, what, and how we'll plant a billion trees by 2028. Find out about the programme and what funding and grants are available. Available at: www.mpi.govt.nz/funding-and-programmes/forestry/one-billion-trees-programme/about-the-one-billion-trees-programme

10 Life and death in the Anthropocene

Introduction

This chapter, *Life and death in the Anthropocene*, is located in the great paradox of the Anthropocene, hovering in the inseparable intertwinement of the mighty Amazon River with its rainforest, and its global regulation of atmospheric cycles. It is the time of the sixth great extinction, 'a new phase in the history of both humankind and of the Earth, when natural forces and human forces became intertwined, so that the fate of one determines the fate of the other' (Zalasiewicz et al., 2010, 2231). As the greatest anthropogenic threat to the commons the world has ever known, it is also the time when the human is blessed with one last chance to become part of the planet. This chapter, *Life and death in the Anthropocene*, gathers up stories that have produced themselves as intact stories waiting to be told for the ending of the river's book. They are all stories that sit in the paradoxical space of the Anthropocene, where life and death intersect, where death produces life and life produces death. They are stories of *the Spot of death; the Flowers of eternal life; Burying gold; Mississippi floods*; and *the Gundungurra creation story*.

> The questions raised by the Anthropocene are intimately connected to reconfiguring our relationship to life on earth, not only to the biological workings of ecosystems but also our cognitive, emotional, perceptual, aesthetic, ideational, and cultural experiences of the prismatic living world of which we are part.
>
> (Degeorges and Oppermann, 2019)

The spot of death

Riverlands, 30 July 2016
There are no words
only the images of Mollie
beloved border collie cross
rescue dog
bloated body floating
claimed by the river as its own
gently placed into riverbank grave

under protective cover
of river stones
beginning and end of the river's book
of life and death in the Anthropocene.

The beginning and end of the river's book

The spot of death gets its power from the death of Mollie the dog. Mollie the dog lost her companion, Mercy, the disabled friend that she made abled by their play. When Mollie and Mercy were alive, they walked their human in the riverlands twice a day, every day, exploring all of their places. Then after Mollie and Mercy were gone, the river began to walk their human, taking me to all of its spots, introducing me to its rocks, river stones, its slow flow, its trees and all of its creatures. The river seeped into me and I seeped into the river as we became together and the river's book was born.

We lived together for a long time, until at 18 years, Mercy's heart lost its ability to pump blood and she crossed to the other side, held in my loving arms. Soon after, Mollie, having lost her best friend, started having little seizures. Then one day, walking in the riverlands, she vanished. Calling, calling, calling, did not bring her back. Notices everywhere with images of a small perfect border collie cross lost dog yielded nothing, until three weeks later, a fisherman called, *I think I have found your dog. She is in the river just near the water pumps of the correctional centre.* We go to the spot of death and there she is, washed into a small inlet, bloated decaying body barely intact, and heavy with the river's water. No longer of the land but of the river.

Burial in the river's lands

Should we leave the river to complete her final disintegration, leave her to join the river's eternal flow? But no, we selfishly want to know where she is, where we can remember her small loving body that had become so much part of us. So we return with tools to make a grave by the river, a marker of the spot of death. The hole is hard to dig, and it needs to be large enough to encompass the full extent of her swollen body, too fragile to bend. Here, so close to the river it is all river stones. We lift them out one by one, big ones, small ones, round ones, rough and smooth, all different shapes and colours, and place them aside until the hole is large enough. Mollie's heavy disintegrating body is lifted ever so gently into its bed of river soil, covered over with dirt, and then with the river stones that were removed to make her grave. Mollie is now forever part of the river's lands at the Spot of death.

The colours of Mollie's sunset

It's hard to leave her there. Tears flow with the river's water as we walk away, scrambling through weeds and bushy undergrowth from the *Spot of Death.*

Then at the top of the hill above Mollie's new burial place, the sky bursts into a shower of vibrant colours. A sunset lights up the whole wide arc of the sky into a brilliance of colour that we've never seen before or since at the riverlands. The rugged dark line of the horizon is broken only by the dark silhouette of two trees illuminated by a burst of golden light. The gold merges into the brilliance of vibrant pink shining across the arc of the sky in puffs of coloured light. Higher up, the vivid pink blends into the darkest ultramarine that turns purple when it mixes with the pink. A single swathe of dark blue cuts across the whole light show, criss-crossing fine power lines that are strangely beautiful, transformed into an element of the sky's artistic creation. We leave feeling deeply blessed by Mollie's parting gift of vibrant cosmic wonder at life and death in the time of human entanglement in the fate of the planet.

Small girls, river stones and flowers

It is some time since our beloved dog disappeared in the vivid sunset sky. Two small girls are still missing Mollie the dog. *Where is Mollie?, they ask.* She's gone to the riverlands, we say, she will always be there in her spot under the river stones that mark her grave. Would you like to go there and see it? Yes, they say, can we take some flowers, and immediately busy themselves picking bright pinky-red camellias, purple daisies, with deep black centres, a sprig of newly opened pale pink spring blossom, sprigs of purple lavender, a cluster of unopened dark maroon magnolia bud and some red petals. We walk to the spot of death where they talk to their little dog friend that they have known all their lives. They sit by the river stones on her grave, small hands lovingly placing each flower, petal, cluster of buds, sprigs of lavender and spring blossom on the stones, as they talk about Mollie. Her grave now vibrantly coloured with flowers that echo the colours of Mollie's sunset, their act of mourning creating profound beauty in their childlike performance of life and death.

> Earthly colors immemorially attune us to life's prismatic richness: chromoscapes and polychrome expressions are an attribute of all life forms, faunal, floral, bacterial, and even mineral. Color dynamics shape the life experiences of humans as well as nonhuman entities, manifesting in the permeable boundaries between them, as color plays a constitutive part in more-than-human environments.
>
> (Degeorges and Oppermann, 2019)

The flowers of eternal life

> ***Drainway, 21 March 2019, 9.30 a.m.***
> Grey misty day gentle light rain
> as burst of yellow flowers lights up
> dull sad mind (they're only weeds).
> Walking on and more and more

flowers burst brightness and life into
this dull sad day.
Purple morning glory
white open-throated flowers
drape over weedy shrubs,
delicate pale pink petals
with deep maroon throat
shine to the new day.
Yellows, pinks, purples, whites, reds,
every colour of the universe
in the eternal riverlands of life.
We are all migrants here.

Flowers for massacre's trauma

It's the day of the Children's Climate Strike, a day when children all over the world protest against the lack of global action to save their planet Earth. It's a day of hope, the one day we decide to watch the news:

> The Christchurch mosque shootings were two consecutive terrorist shooting attacks at mosques in Christchurch, New Zealand, during Friday Prayer on 15 March 2019. The attacks began at the Al Noor Mosque and continued at the Linwood Islamic Centre. The attacks killed 51 people and injured 49. The suspect [no name] was a 28-year-old Australian man, described in media reports as a white supremacist and part of the alt-right, was arrested and later charged with 51 murders, 40 attempted murders, and engaging in a terrorist act.
>
> (*ABC News*, 2019)

Tears, shock, trauma. I am unable to accommodate this act of hate that seems to become irrevocably connected to the Children's Climate Strike about the end of the world as they know it. I cannot understand how we have come to produce a world where the sacred becomes the reason to commit such violence. Where innocent people are killed, where a whole country and community are devastated by the trauma of mass shootings.

In the hours and days that follow, there is some relief in the responses of leaders all over the world and especially the New Zealand prime minister, Jacinda Arden, who immediately mobilises the country to condemn the crime and the white supremacist sentiments that make it possible. The mosque is overwhelmed with expressions of sympathy. All over the world, people express their grief, love, sorrow and support, by laying flowers at the sacred mosques.

I need to act, we must lay flowers at our university, for our large numbers of Muslim students and staff. I imagine a line of flowers along a retaining wall that stretches some hundreds of metres from the library towards the childcare centre. I imagine they would be laid informally by anyone who wants to express their sadness, and show love and kindness to our Muslim colleagues. Due to the university's red tape, we end up with a small ceremony that I am asked to lead:

> We invite all staff and students of Western Sydney University's Kingswood campus to place a bunch of flowers on the lawns outside the prayer house at the back of Building J-K in an expression of our great sorrow at the hate crime carried out in New Zealand. These can be modest flowers from your own garden, or bought from a local florist where we convey our intentions. We do this to express our loving compassion for all those affected directly and indirectly, to publicly represent the values of loving peace and harmony that we would like to promote in the world, and especially to signal our university's commitment through all of our teaching, learning, research, and service activities to continue to build the kind of society that enacts this loving compassion in our everyday lives.

It's a grey misty day of light rain as I walk, feeling the sadness and trauma of so much suffering, and then the bright colours of flowers insert themselves into my consciousness. The golden glow of cassia bush flowers shines light into the dull sad day. Turning the corner on my usual walk by the river, I am confronted by a vine with delicate white flowers garlanded over an impenetrable pathway of weeds. Further on, a purple-flowered morning glory vine is draped over taller woody shrubs. After that, the delicate pink of a wild hibiscus weed is just opening to the morning's light. I marvel at the fact that I have never noticed these flowers before, the brightly coloured flowers of weeds somehow invisible. On this day, the flowers arrange themselves in my mind as a bright bunch of glorious colours, a precious gift of the riverlands for a flower-laying ceremony. I wonder what it is that makes these flowers so brightly coloured in their own artful excess of the world's creation:

> Colour exists solely to reproduce and procreate. The colour of flowers are found in pigments that are decided upon in the hereditary genome of the plant. Flower colours of red, pink, blue and purple come mainly from the pigments called anthocyanins in the class of chemicals called flavanoids. Other pigments are carotenoids, that provide yellow, red and orange pigments in the plastid cells of plants. Flowers that are bright in colour are meant to attract birds, bees and other insects in order to help the plants reproduce. If a plant needs to reproduce with the help of the birds and the bees, the genetics will make the flowers have bright colours.
>
> (Florapedia blog, n.d.)

Harvesting flowers for ceremony

It's the morning of the ceremony. I wake early to a bleeding sunrise sky with long clouds lit vivid pink and intermixed with blues and purples of deep rain clouds. I set off to the flowers at riverlands, tools and shopping bag ready, wondering about how to harvest the flowers, arrange them, transport them to the university intact, and in time to get the ceremony organised. I walk fast in the early light, breathing out hard as if breathing under the water, releasing body tension, breathe in, breathe out, when the bright red flowers of the coral tree, laden with bees interrupt my reverie. I cut some with secateurs, careful to avoid their sharp spikes, place them in a cool pack shopping bag and walk on to the twining white-flowered vine, along the impenetrable weedy alleyway next to the drain. It is impossible to de-twine them from their weedy entanglements so I cut whole structures of weedy stems intertwined with white, open-throated flowers touched with the faintest strokes of purple.

In the deep shade, the morning glory isn't awake yet, its petals still closed in sleep. On the sunny side there are more vines, decked in deep pink flowers that invite clambering in to reach those already open towards the morning. Further along, I harvest purple ground weeds with a pungent perfume but cannot find the wild hibiscus until suddenly behind the ridge of wild fennel weed, a small branched structure is decked with pale pink buds still almost closed in their night-time pose. I clip the stem and add to the growing pile of flowers damp from much-needed rain in this old dry land. The motley bunch of mixed up flowers, vines, weeds and creepers accompanies me home, falling out of the cool bag, and placed in the sun for the final flowers to open before arranging in their ceremonial postures.

I gather the flowers into a loose bunch, entwining the white, purple and pink flowering vines over the golds, reds, pinks, and purples of the others to express their sense of wildness, and load them into the car with the bunch I have bought from our local Woolworths and another from the back garden.

Flower-laying ceremony

We gather together in the early morning outside the prayer house to express our mourning at the profound losses we have witnessed, and our commitment to loving kindness, peace and harmony. A number of young Muslim students come with their lecturer, traumatised yet strong in expressing their combined sadness, fear and gratitude. Tears are shared from all of us, and each person speaks as they lay their flowers and listen in loving silence to everyone's flower-laying words. The director of the childcare centre lays flowers from her staff, and a wreath of flowers made by the children from the shape of their hands. We can hear their joyful voices in the background as their wreath is laid. As I lay my three bunches of flowers, I shape words of their meanings, words given by the river as I harvested its bright colours, *The flowers from my river represent healing from trauma, they are flowers from the rivers of eternal life.*

An artist donates a beautiful painting of the power of love. The painting is a glory of colour echoing the river's flowers: hearts in layered bright red, dark red, pink, and yellow intertwined with winding white dots. The hearts are enmeshed in a massed wildness of colour: red flowers, yellow flowers, pink flowers, blue flowers, orange flowers, imaged against a vivid ultramarine blue. Echoing vibrant colour: the colours of Mollie's sunset, the colours of the flowers on Mollie's grave, the colours of the river's wild weed flowers, the colour that originates from the plants of our worlds.

After our flower-laying is over, we retire to the café to talk together and share our reflections on our ceremony. One by one, people join us and leave until, serendipitously, after several hours of sharing, relaxing and talking, two of us leave the building. At this very moment our Muslim colleagues are filing out of the prayer house from the opposite direction. We meet each other on the pathway. First, the leader shakes our hands and speaks of how much this means to his people and how recovery and healing from trauma are so vital. Then a young man and woman who wait behind, respectfully, for the leader to finish, approach us, ask if they can hold our hands, (together we can), and with a mixture of both tears and joy, tell us in halting words, how important the flower-laying was to them, how it helps them to heal from the trauma, and makes a difference to how they feel about living in this country.

Over the next days and weeks the flowers, placed in among the bushes in the garden slowly fade and die and I watch the river's flowers of eternal life little by little, day by day, lose their glorious colour, turn brown and dry and eventually disappear into the earth from whence they came.

> We can tune into the drama and evolution of life through colors to see how connections on our symbiotic planet unfold in more ways than one. This interpretation of the earth (and of life itself) helps ecological imagination recast the color spectrum into a new understanding of biodiversity.
>
> (Degeorges and Oppermann, 2019)

Burying gold

My doctor tells me everybody has trauma, but the question is what do we do, what can we do with our trauma? What can we do with the trauma of the Anthropocene? I have a golden cloak that symbolises trauma for me. It was given to me but I cannot wear it, it is full of the trauma it represents. I do not know what to do with it. My counsellor suggests burning it, but I do not want to burn it, it feels too violent. My friend suggests giving it away to someone else but that doesn't work, it only passes the trauma on to another person. My daughter suggests I give it to St Vinnies op shop but I do not want the trauma to exist in the world. After much contemplation I make a plan to bury the trauma.

Binding gold

Today is the day. I take the golden cloak with its glorious colours from the drawer where it has remained hidden and it immediately escapes, the heavy crinkled gold with its vibrant colours unfolds, spreads uncontainably. I cannot wait to contain its iridescent flows. I start to fold it, and then roll it up but it springs out, it will not stay in place, so it becomes a process of *binding gold*. I start with rubber bands, easy to place as I hold the folded roll with one hand and slip a rubber band on with the other. Halfway secured, I tie it with gold ribbon, at intervals along the roll, tying it in tight knots to hold it in place. It is still unravelling, escaping its ties in tufts of gold pink red blue and purple. So I bind it, winding the golden ribbon along its whole length, starting with a tie and one end, round and round and round I bind, pulling it tight all the way along its roll until I come to the other end and tie it off, to seal the binding.

I fall in love with this artwork that is now produced, it is beautiful, the brilliant colours interwoven with gold still visible through the gold binding that holds it tightly in place, the pain and trauma are now tightly contained within. It is a beautiful trauma pupa, waiting to metamorphose into its final form, it is a chrysalis, the quiescent form of the pupa, sleeping until it is ready to birth something new. But I am done with this stage of binding gold. My neck is sore, body tired from the symbolic pain of a lifetime, but soon it will be gone. I put it back in its hidden place until I am ready to take my glorious bound golden paradox of trauma to join my beloved river at Mollie's spot, the spot of death.

Riverlands, 13 June 2019, 3.00 p.m.
It's time
gold leads the way
a cluster of baby gold mushrooms
spring from the base of a tree
large plate-size orange gold
toadstools at the bottom of
a dead stump
overhead shining gold leaves
of white cedar tree reach
into deep blue sky
and shower down in light wind.

It's a day of golden light. I dress in walking clothes, old familiar blue overalls. My golden chrysalis is placed in the bottom of a silver-lined blue Aldi cool bag, a bottle of water, swimmers and towel, and head off to the prickly pears that mark the way down to the spot of death. I find the parting of the grasses, the walking trail, that I last walked with three small girls. I follow the faint storyline trail, winding through long grassed, rugged terrain until I reach the river where I turn towards the spot of death. But I cannot get there, caught in trailing balloon vine, prickled with powerful coral tree thorns, blocked by fallen lantana foliage and a barbed wire fence.

Riverlands, 13 June 2019, 3.55 p.m.
Dark green water alive with
black swans tiny grebes
red-beaked moor hens
that flutter and float away upstream
landing near Mollie's
spot of death.
Lean against comforting trunk
of casuarina at water's edge
in light high-pitched chorus of water birds
deep caw caw of crows
swishing wind of casuarinas
until the pick calls to me
it is time.

Finally, hot and bothered after much struggle, I arrive at the river to begin my lone ritual. But I am so not alone. As I arrive at the calm green river, hordes of water birds scatter and flit, flicking the still surface, squealing, cooing, and splashing. Flocks of majestic black swans, tiny busy grebes, red-beaked moor hens, cormorants, and a lone pelican. They continue to watch as I move about, a comforting riverine presence.

After a while of sitting, catching my breath, I decide to begin digging right there under the casuarina tree, but it is roots, roots, and more roots, digging is not possible. So I find another spot that I think is next to a wooden pole but soon discover it is a sawn off casuarina tree with worm holes and flaking bark. I dig there anyway. The pick loosens up damp sweet-smelling soil, crunches through roots, and hits rock. The first one, large and deep, but after levering it out, it opens the way for a long oval-shaped hole, with more digging and scraping and scratching out stones and rocks. Chomp, dig, level, and scrape over and over again until its done. There is a hole big enough to lie my golden chrysalis in.

I need a pause and I find some casuarina needles to line the oval hole of river earth for my chrysalis. I slide back to the casuarina by the river, soft needles gathered at its base, and scoop them up to make a nest for my folded and rolled gold. I want to make it nestled and clean. The soft fine needles are spread along the hole. I place the gold-bound pupa in its nested container. Its gold and vibrant colours shine against the grey brown needles echoing the colours of Mollie's sunset, the flowers the little girls placed at this spot of death, and the river's flowers of eternal life.

Pause again, I need more needles, slow down, back to casuarina, now swishing in the wind calling all the names of all who ever walked here as the water birds call theirs to the silent river. In that moment I am suddenly with Bedrock's gold, the two intermingled, the gold of healing, and the gold of pain and trauma. The first shining gold light that became my spiritual meditation, and the second miracle of gold's play of sunlight, water and rock. Deeply calm, I move back to the nest of gold and place more needles gently over the top so that now it is hidden from sight. A slight loss but a feeling of wonder about what might come next now the gold chrysalis of pain is concealed and gentled in its nest of needles.

It is time for the earth, the sweet-smelling, dark brown, river earth to be placed over the nested gold. I scrape it in from the edges of the burial hole, covering it gently so as not to force the dirt in through the bed of needles. Then suddenly I am crying, sobbing for the loss, the pain and trauma, the small girl child that was me. It is within and without, buried with the gold but still in the child within that is me. I cry and sob some more, then find myself patting the dirt, and realise it is the tapping of trauma work that brings memories to the fore from somewhere in the unconscious brain. So much of trauma work is unconscious body work. Tap, tap, tap, go my hands, patting dark earth until it is smooth and even across the hole. Tap, tap, tap, hands go and then it stops, body is stilled.

Hands form around river rocks, big ones and small ones holding them, feeling their rounded certainty, they are my bedrocks, healing rocks, formed by the river water's constant movement against the lithic materiality of rock. The river stones are soothing, as beautiful as gold in their neat patterns of long, round, flat, big and small, gentle river stone colours. They form a protective cover on top of the buried gold, sealing in trauma, giving it a last shelter where the gold chrysalis can become something new.

Thinking the subject as a material being constrained by the agencies of the compromised, entangled world enacts an environmental posthumanism, insisting that we, as bodies and minds, are inextricably interlinked with circulating substances, materialities and forces. The relational capacity of the posthuman subject is not confined within our species, but it includes all non-anthropocentric elements, starting with the air we breathe. Living matter, including human flesh, is intelligent and self-organising, but it is so precisely because it is not disconnected from the rest of organic life, the animal and to the earth (Alaimo, 2017; Braidotti, 2017).

Beginning and ending with water #1: Mississippi floods

19/04/19

Dear Emu Girrl,

Last night I spoke to my mother about the floods in Nebraska. Every time we talk there are new flood updates. Stories materializing like the flood waters. As these words passed through phone lines, I turned to the materialities of the river. I think of how much concrete is pouring through the rivers. There are so many bridges and roads that the waters have ruined. For me to drive home, it would take an extra 2–3 hours because of washed-out roads and bridges. She told me of family heirlooms. The water took with it a confirmation dress that was worn by 5 generations of women in a family. The water also took family photographs. Washed away decks from homes that anticipated summer parties and sittings. Toys that little hands still grab for in play. Material pasts taken up by the river, disintegrating them into unknown futures. The rivers' reach extends out, grabbing what it can, relentless as it moves outside of its borders.

FS

02/08/19

Emu Girrl,

The images of the shoe and tree were taken in a location that had a sign in front of it warning of possible danger. Still. Months after. I manoevered over contaminated sand, sand and sediment from the river that escaped from the river, formed piles, and made new creations. Skeletons of animals partially buried in sand. Debris from homes. A shoe-earth-river creation. The river giving objects new homes. Tiny frogs hopping with each of my steps. The trees with their barks stripped by icebergs exposing human folly, the folly that we have some kind of control while we're here. We're just passing through, hoping not to strip the bark from trees. I do hope that you had some lovely time with your beautiful Nepean. Rest so that you may flow.

FS

Recognizing that we live in a geological moment that allows us to observe the earth "tearing down, dancing over, laughing at" our efforts to restrain it, we geologists are trying to put this recognition into words. But we are embedded in controversy swirling around the proposal to acknowledge a new geological epoch.

(Schneiderman, 2017, 170)

Beginning and ending with water #2: Community update, Water NSW, Summer 2018/2019

A pamphlet appears in my letter box with a letter informing me that my property is in imminent danger of flooding. I remember a long time ago hearing the roar of the river after the heavy rains of an east coast low, but even then the river remained within its banks. Flooding seems so improbable in this time of drought that I read on:

> The NSW government proposal to raise the Warragamba Dam wall to create a flood mitigation zone of around 14 metres is a key element of an integrated strategy to reduce significant existing flood risk in the Hawkesbury Nepean Valley. This Community Update shares information about how the proposed project will better protect downstream communities from threat to life, safeguard property and maintain essential infrastructure. Investigations are now underway for the proposal to raise Warragamba Dam so that it can be used to capture the peak of a flood and release it gradually and to better understand the upstream impacts of the proposal.
>
> (Water NSW, 2018)

The map on the first page shows a vast area of green wilderness penetrated by winding blue waterways upstream of the dam. Each blue waterway has curly red around its edges, depicting the possible flooding to be expected in the Burragorang Valley upstream of the dam: 'If the dam wall is raised, preliminary modelling indicates a temporary inundation increase in the Greater Blue Mountains Heritage Area of 0.04–0.05% of the total area.' Doesn't sound much, something equivalent to the minor effect of the curly red lines. But I read on and it equates to an area of 400–550 hectares. I find a link to the Colong Foundation 'Give a Dam Campaign' website so I can check out what all this means to life and death in the Anthropocene.

Give a Dam campaign

> This is where our people come from, so much could be destroyed by one flooding. episode.
>
> (Aunty Sharyn Halls)

Desecrating Indigenous culture

The southern Blue Mountains is an extensive and rich cultural landscape belonging to the Gundungurra people. The rivers, waterholes and mountains of the Blue Mountains landscape tell one of the most intact and documented Dreamtime stories in Australia – the epic battle of the tiger cat (Mirrigan) and the snake (Gurrangatch) which formed the southern Blue Mountains.

When Warragamba Dam was built in 1960, it resulted in the flooding of a large proportion of the cultural heritage and Dreamtime stories of the Gundungurra people. If the dam wall is raised, the remaining sites of this story – including Indigenous archaeological sites, creation waterholes, and cave art – will be destroyed.

They say the raising of the dam wall is being driven by developer interests on the Hawkesbury-Nepean floodplain to allow an additional 134,000 people to reside on western Sydney floodplains after the dam is raised.

An area of 4,700 hectares of the World Heritage-listed Blue Mountains National Parks and 65 kilometres of wilderness streams would be inundated by the 14-metre dam wall raising.

The wild rivers of the southern Blue Mountains are home to 48 threatened plant and animal species, ancient river valleys, rare dry rainforests and hundreds of Indigenous cultural sites.

The Give a Dam campaign produces a film about the Burragorang valley and the Gundangara people's struggle to save their sacred songline. The film is shown at Warragamba, the place of the Dam. The place where my great-uncle, the mystery Wee Davy, and his son, laboured to build the largest dam in the world at the time, now the place of the water supply of the mega city of Sydney. The dam that has already submerged: 'Much of this landscape with its minute Gandangara toponymic descriptions considered to be one of the best documented Aboriginal cultural landscapes.' I feel deeply implicated in this paradoxical place, as if part of my DNA is embedded in the dam, and in the waterways of the Burragorang Valley.

A songline lives deep within the valley

The Give a Dam campaign launches its campaign to a large audience of both locals and protesters, at the Warragamba RSL club. I have never been to the small village of Warragamba, but feel connected to the place itself though my past and living relatives who may even be present. The film begins and I am lost to these musings, transformed into Gundangara Country.

We are floating over a vast mountainous landscape, blue green mountains folded into deep valleys amidst floating mists, white puff clouds, and eery atmospheric music. A young woman's voice emerges from these misty mountains, shining out loud and crystal-clear: '*Yarangi yangunburra manduginni ganugurra nunga yanangi:* On behalf of my people we welcome you to Gundangara lands.' Hers is the voice of Country, the deep mountain valleys of the Gundangara peoples.

Suddenly the eerie music stops and we are transported to a room of padded leather seats and railings, filled with men dressed in suits and ties. A large dominant suit and tie voice steps forward and pronounces:

> Mr Deputy President I move that leave be given to bring in a bill for an Act to make provision with an amendment to the NSW Act of 2014 in respect to the temporary inundation of national park land resulting from the raising of the wall of the Warragamba Dam for downstream flood mitigation purposes.

As soon as he finishes speaking, we enter into the deep gorge of the Burragorang valley, down down we go, winding with its waterway through rocky cliffs, high mountain escarpments, cascading over rocks and stones formed by the gentle movement of water on the lithic materiality rock over millions of years.

And then again we are faced with the room in the NSW Parliament where the voices of interested parties plea for political sensibility to the needs of the waterway, to everything that lives in the Burragorang valley, and to the deep time heritage of its cultural stories.

The film moves in and out of these scenes with much of it moving through the deep valley of Gundangara Country, the place of the proposed 'temporary flooding'. We dwell there with its animals, a snake swims through the water, black cockatoos squeal noisily as they fly overhead, kangaroos hop through the grasslands. Strolling through a rare remnant stand of old growth forest, we are surrounded by birds of all different colours, sizes and shapes, singing, calling, nesting, flying from tree to tree. But especially we get to know the endangered regent honeyeater, if these trees are flooded, its habitat will be lost and they will die:

> The Regent Honeyeater is a striking and distinctive, medium-sized, black and yellow honeyeater with a sturdy, curved bill. Its head, neck, throat, upper breast and bill are black and the back and lower breast are pale

> lemon in colour with a black scalloped pattern. Its flight and tail feathers are edged with bright yellow. There is a characteristic patch of dark pink or cream-coloured facial-skin around the eye. The call is a soft metallic bell-like song; birds are most vocal in non-breeding season.
>
> (NSW Department of Planning, Industry and Environment, n.d.)

The language of water is everywhere, water bubbles over rocks, roars over cliffs, tinkles with droplets of rain, or lies still in silent pools. I flow with the water that lives deep within Gundagurra Country to the songline of its creation, a songline that can only be properly known in Country, a disrupted songline where 11 of its 15 sacred waterholes are already submerged beneath the waters of the dam:

> In *gun-yung-ga-lung* the primordial Dreamtime, two creator ancestors, *Gurangatch*, a rainbow serpent and *Mirragañ*, a quoll, went on a journey from a point on the upper reaches of the Wollondilly River, with Mirragan pursuing *Gurangatch*, until the chase ended at a waterhole named Joolundoo on the Upper Fish River. The distance covered by this serpentine movement and the pursuit extended some 169 kilometres. *Gurangatch*, not wholly a serpent, but part fish, and part reptile, camped in the shallows of an area known as *Murraural*, at the junction of the Wollondilly and Wingeecaribbee rivers. It was here, while he basked in the sun, that the fish-hunter, Mirragañ the quoll, glimpsed the light reflected from Gurangatch's eyes and endeavoured, unsuccessfully, to spear him. The quoll tried to force his prey back from the depths of the waterhole, where Gurangatch had sought refuge, by planting ever more bundles of nauseating slabs of *millewa* (Hickory hickory bark) here and there in the various soaks and pools. Gurangatch, wise to the plan, burrowed his way out, tunnelling through the landscape, drawing the lagoon waters in his train, till he emerged on a high rocky ridge called thereafter *Birrimbunnungalai*, since it is rich in sprats, small oily fish known as *birrimbunnunnung*.
>
> (Gundungurra Aboriginal Heritage Association, 2018)

> The features of the landscape were etched as Gurangatch wriggled and slipped across and under the terrain, fleeing from his predator, or sometimes while directly fighting with him. When Mirragañ caught up with his prey, he would flail away at him with a club (*boodee*), while Gurangatch would strike by thrashing his tormentor with a whipping from his tail. This place is *Wonggaree*, now known as *Slippery Rock*. It marks a point where they engaged in struggle for a long time, wearing the rock down so smoothly that people slip on it ever since. It is a marker in the landscape of its creation, a sacred place whose story remains in the last remnants of unflooded waterways deep within the hidden valley.
>
> (ibid.)

For its people, the *Gandangarra, Gundungurra, Gundungari, Gundanora, Gurragunga* and *Burragorang, the story only truly lives in language and in Country.* Its story waits for the outcomes of political will; if its Country is flooded, both story and its language will be lost. The power of singing the Country into being in sacred ceremony and songline over and over again since time immemorial, will be gone forever. And yet the song lives on. Wherever I search, there seems to be another version of this story, all slightly different in their living oral presence. The whole songline is mapped onto Country in a submission presented by the Gundungurra Aboriginal Heritage Association to the parliamentary inquiry. The journey of Gurangatch and Mirragan is the central focus of the submission. The document includes nine maps of the now familiar winding waterways of the Burragorang Valley. The first displays the entire creation story, and the remaining eight maps illustrate further detail of the story's places. There are 57 individual sites described in minute material and narrative detail.

I traverse the maps over and over, dwelling in the intimate detail of the creation story of the flooded Burragorang Valley. Every story place mapped and known in all its intimacy. There is a proliferation of language names that my mouth longs to learn the feel of, names like *Murraural* Waterhole; *Birrimbunnungalai* Waterhole; *Gunggalook* Waterhole; Belloon Pass *(Belonong); Woonggaree* Waterhole; *Goorit* Waterhole, all part of the magical ongoing cultural knowledge of the Burragorang Valley. I learn that some of them are marked with art works: Waratah Rock has hand stencils, a dancing figure, and an imprint in the shape of a waratah flower. There are medicine places with *Morle Boc* demonstrated to contain natural antibiotics and connected to the Gundungurra Ancestor called *Gareem*, the God of sickness and health.

There are story places of other animals caught up in the long long journey of Gurangatch and Mirragan as they created the landscape of this country: *Cunnark* Waterhole (mud or black eel); *Warrumba* (eastern long-necked tortoise); *Billagoola* Waterhole; (the black shag or great cormorant, from *billa* the Belar tree; *Casuarina cunninghamii*; and *boola*, water birds).

Finally, the storyline ends at *Joolundoo* Waterhole; the resting/hiding place of Gurangatch:

> The exhausted Gurangatch found relatives in the dark depths of Binnoomur (Jenolan Caves) and they took him to a deep waterhole, Joolundoo. There they hid him with their smaller bodies, and his head and tail became wedged in rock crevices. Mirragan, knowing where Gurangatch was hiding, called on the best divers to assist him – Billagoola the shag, Goolagwangan the musk duck, Gundhareen the black duck and Goodarring the wood duck. Each dived down into the waterhole surfacing with one of Gurangatch's smaller relatives, but finally Billagoola reached Gurangatch. He could not shift him but managed to tear off a great piece of flesh from his back. The hunter and divers cooked the meat, shared 'a great feast' together and then returned to their homes.
>
> (Karskens, 2010, 26)

I long to go to these places, the ones that remain still intact from the flooding, but I cannot, there is still more to do at my river.

Last visits

Last visit #1, 10 August 2019, 10 a.m.
Gusty cold windy day
hint of spring in wattle blossom
wind down rugged pathway
to spot of death
lost in multiple trails
wind gusts forcing backward
stuck in prickle bushes
stumbling on balloon vine
finally arrive but
there are no stones
carefully placed in patterns
of different colours and sizes
all gone.
Leave via tree stumps, stable solid not going anywhere.

Last visit #2, 11 August 2019, 8.30 a.m.
Compelled to find buried gold
blue overalls walking boots
find the way through bumpy grass
to barbed wire fence
down over river bank through
privet's low-hanging purple berries.
No surprise just same soft layer
of casuarina needles
beside sawn-off trunk with flaking bark.
Clear away fine needles
scrape into sweet-smelling river dirt
to big river stones
immoveable with no tools.
Rip and scrape rip and scrape still no more
little rocks of all shapes sizes colours
only strip of clear plastic
Sufix Surfix Sufix in silver outlined in black
and wide open fish mouth layered with other fishy bodies.
Leave via tree stumps.

Last visit #3 13 August 2019, 3.39 p.m.

There is one more call to walk the songline of burying gold and the spot of death, to be with the beloved dog that the river took for its own, the colours of sunset, and small girl's flowers of mourning on the river's stones. The flowers of the rivers of eternal life, all the colours of the universe harvested and laid with love, all weeds, all migrants in this place. The colours of the gold cloak of buried trauma, a beautiful chrysalis, quiescent, waiting to re-emerge as something new.

This time walking the river's flood plains holds its memory traces of all that was lost in the time of the great floods, houses and trees, human lives, all swept away in the power of the river's crashing torrents. The steep bank that I clamber down through low-hanging branches with purple privet berries is the wall that marks the riverbank before Warragamba was built. The time before Warragamba Dam was built, the dam that flooded the songlines of the Burragorang Valley, the dam that was built by my great-uncle, the mystery Wee Davy. Everything resonates with connections and memory traces.

As I sit under the casuarina by the river, I can hear the river's flow rippling and gurgling over rock, the same water, the same flow, that comes from the place of the sacred songline of Gurangatch and Mirragan, the river's creation stories, connected now by the river's flow at this place of burying gold. A red-beaked moor hen calls and a lone pelican and cormorant sit on a log in the wide pool where small girl voices rang out in their lullaby to seal babies. Lonely caw caw of crows reminds of the long slow winter of healing and the miracle power of gold in the dance of sun on water and rock. The gold of healing and the gold of trauma intertwined. It's time.

This time I have brought the blue Aldi cool bag with gloves, a hand fork, and water bottle, so I begin the work with some trepidation not knowing what I will find nor how it will affect me. Once again I scrape away the soft casuarina needles and dig through the roots, ripping and pulling to remove the sweet-smelling river earth. I reach the large oblong river stones, smooth and solid, and lever them out with the fork, placing them alongside the growing hole. Then underneath are the lost river stones, the ones that were on top in all their different colours as sizes, as beautiful as gold in their own way. Digging and digging, I remove them one by one, down to the smallest until they sit by the hole next to the big rocks. One last one is small, flat and gold-coloured. Finished now, there is no cloak to be found here, I take off my gloves and feel the smoothness of this small river stone. I place it in the pocket of the blue overalls, a last memento of the buried gold chrysalis, a smooth gold stone with a small imperfection on one side, the scar of loss.

The empty space of buried gold returns me to that place over and over again. The empty hole of absence and loss creates a space for a new storyline, a story of connections, a sacred place where multiple storylines intersect, a place of singing Country, and all who dwell there, into well-being. I will never know where buried gold has gone, just as there are no answers to the Amazon's question of life and death in the Anthropocene, hovering in the intertwined space of the river and its rainforest:

> A work is supposed to bring out problems and questions in which we find ourselves caught, and not to provide answers. A work is a new syntax, which is much more important than diction. It carves out a foreign language in the language.
>
> (Deleuze, 2006)

References

ABC News (2019). Report of NZ massacre. 15 March. Available at: www.abc.net.au/news/2019-03-15/christchurch-shooting-multiple-fatalities-mosque-new-zealand/10904416

Alaimo, S. (2017). Your shell on acid: material immersion, Anthropocene dissolves,, In R. Grusin (ed.), *Anthropocene Feminism*. Minneapolis, MN: University of Minnesota Press, pp. 89–121

Braidotti, R. (2017). Four theses on posthuman feminism. In R. Grusin (ed.), *Anthropocene Feminism*. Minneapolis, MN: University of Minnesota Press, pp. 21–48.

Degeorges, P. and Oppermann, S. (2019). Reclaiming the entangled colors of life in the face of the Anthropocene. *BifrostOnline*. Available at: https://bifrostonline.org/reclaiming-the-entangled-colors-of-life-in-the-face-of-the-anthropocene/

Deleuze, G. (2006). *Two Regimes of Madness*, ed. D. Lapoujade, trans. A. Hodges and M. Taormina. New York: Semiotext(e).

Florapedia blog (n.d.). Available at: www.proflowers.com/blog/category/florapedia

Give a Dam Campaign (2019). Available at: www.giveadam.org.au/film_video

Grusin, R. (ed.) (2017). *Anthropocene Feminism*. Minneapolis, MN: University of Minnesota Press.

Gundungurra Aboriginal Heritage Association (2018). Inquiry into Water NSW amendment (Warragamba Dam) Bill, 2018. Available at: www.parliament.nsw.gov.au/lcdocs/submissions/62428/072%20-%20Gundungurra%20Aboriginal%20Heritage%20Association_Redacted.pdf

Karskens, G. (2010). *The Colony: A History of Early Sydney*, Crows Nest, NSW: Allen & Unwin.

NSW Department of Planning, Industry and Environment (n.d.). Regent honeyeater: profile. Available at: www.environment.nsw.gov.au/threatenedspeciesapp/profile.aspx?id=10841

Schniederman, J.S. (2017). The Anthropocene controversy. In R. Grusin (ed.), *Anthropocene Feminism*. Minneapolis, MN: University of Minnesota Press, pp.169–189

Water NSW (2018). Community update. Sydney: Water NSW. Available at: www.waternsw.com.au

Epilogue

Song of mourning from
the graveyard of the world.
Blue green smoke from
monster fire hovers silent
on the ridge
ground beneath my feet
crackles with dry grass
dust bursts into air
from threadbare soil
charred eucalypt leaves drop
from fire-weary sky
curled body of fox lies
dead in dry creek bed
green wattle blooms in pungent
reminder of young fathers killed
in Green Wattle Creek Fire
as we mourn for Planet Earth.

I write this with lungs that cannot breathe the air, as blackened gum leaves drop from the sky, waking daily to the smell of smoke, comforting children who have been evacuated, and others who have lost acres of forest with houses saved by members of their local communities who are volunteer fire fighters, witnessing an (e)motion in Australia that I have never seen before as we realise that our leaders do not care, cannot care and have no wish for change in climate policy despite the fact that over 15 million acres of land have now been burnt, 2,500 buildings destroyed, at least 27 humans have lost their lives, and an estimated 1.25 billion wild animals are gone forever. Daily I wake to the smell of smoke coming from the monster fires. No one is left untouched as the ancient Gondwana forest of the Lapstone monocline burns, and the songlines in the Burragoorang Valley are incinerated.

The season of fire gathers all that has gone before it in the *Riverlands of the Anthropocene*: the ultimate importance of water; the regenerative power of trees in this old dry continent that has evolved with fire in deep time; the necessity of understanding not only Aboriginal ways of dealing with fire but deep Aboriginal ontologies within which fire is just an integral part. The healing that is now required from the trauma of fires through the power of witnessing regeneration; the vital lifegiving gift of the river's water.

Gudalong Marool

It is the D'harwal season of *Gudalong Marool*, hot and dry, the breeding season of *Burran*, kangaroos and wallabies. The blooming of *weetjellan*, the hickory wattle acacia is the major plant indicator of the arrival of the season of *Burran*. The *Wattun'goori* (*banksia serrata*), the witchetty grub tree flowers, the kurrajong trees from previous year are ready to be harvested, the fruit of pokulbi (*dianella caerulia*) ripens, the largest tuber of Gamarral is harvested. the annual season *Gudalong Marool* intersects with the 10–12 year *Mudong* Cycle (Bodkin, 2017).

Djuli of the *Mudong* cycle

The *Mudong* cycle is a decadal cycle that marks the major songlines and travelling pathways for ceremony. During the time of *Djuli* of the *Mudong* cycle, wildfires were expected and the nests of *Mugatjin*, the black bull ants (*Leptomyrmex spp*) were examined daily. Several days before a violent firestorm, the bull ants start to scurry feverishly up trees, preferably the *Burringoa* or *Eucalyptus tereticornis*, gathering food. When the entry to the nest was sealed and not a single *Mugatjin* could be found, then preparation was made by the clans to leave and seek safety in another place preferably at the base of a cliff facing away from the wind direction … or in specially dug holes under large rocks.

> These fires are unpredictable and dangerous usually caused by lightning strikes or by spontaneous combustion when there is a combination of extremely high temperatures, low humidity, gusting north westerly winds and of stringybark and smooth bark gum trees growing in close proximity … The people made their way to creek or river beds, billabongs, swamplands, or if possible to the coast, awaiting the weather change that indicated that conditions were safe to return to their usual homes … Fighting such fires was considered to be too dangerous, and thus they were left to burn themselves out.
>
> (ibid.)

Everyday wildfire prevention methods

According to Bodkin:

> There were occasions when branches from certain trees were allowed to be removed. These occurred when a certain species of Eucalypt grew in close proximity, so close in fact that their branches were interwoven, for instance, when Bai'yali (Stringybark) grew near any of the smooth barked Eucalypts. The signs of danger to life and land was recognised by the

> D'harawals when a hot dry day occurred and the salt of one's sweat dried and formed a white crust on the skin, the north westerly wind blew in violent gusts causing the interwoven branches to touch and rub against each other. The trees would explode as the bark heat reached its maximum tolerance, and thus, to prevent an uncontrollable fire occurring, the offending branches were removed whenever they were observed before those dangerous conditions occurred.
>
> (ibid.)

The seasons of regeneration

In this season of wildfires at the intersection of *Gudalong Marool*, hot and dry, and *Tjuili*, the *Mudong* season of wildfires, we take refuge from the smoke and heat in the cool of a native plant nursery, walking between its air-cleansing trees for a brief respite and carefully select 5 Eucalypt trees, 2 yellow box, 2 grey box and 1 red box, to plant for the two young firefighters who have lost their lives and the three in a coma with third degree burns. It is the only way we can think of to express our profound grief for so much loss and desolation. We imagine how these trees will grow tall in the riverlands and study their growing habits.

We learn that yellow box, *Eucalyptus melliodora*, will grow to a height of 30 metres and forms a lignotuber which means it will regrow after fire. It's the honey tree, the best native tree for honey production. Red box, *Eucalyptus polyanthemos*, will grow to a height of 20 metres with fibrous or flaky bark on the trunk and larger branches and smooth, mottled, greyish, cream-coloured and yellow bark above. Its leaves will stay green to bluish grey, and provide food for the larvae of the long-tailed bombyx moth and the eucalyptus leaf mining sawfly. Large, old red box form hollows that are used as nests for owls. The grey box or *Eucalyptus microcarpa* is listed as an endangered species. It will also grow tall and strong, up to 25 metres with rough, fibrous or flaky bark on the trunk. Its flowers will be a food source for sugar gliders, squirrel gliders, native birds and insects and its hollows are nesting and refuge sites for native birds and mammals. The grey box will provide critical habitat for the grey-crowned babbler, stone curlew and goannas.

These trees that will grow strong and tall to become a precious gift to the proliferation of life in the river's lands are tiny, delicate, raised in pots and cared for in a nursery. I come to love them through the intimate tending over the days and weeks that follow.

The seasons of planting and regeneration of hope

Day#1: Planting day

It is already hot when we ferry the trees down to the site near the river's slow flow and choose a spot where they will grow as part of a colonnade of Eucalypts to forever mark this time of great loss and the regeneration of hope. I dig with a mattock between dense tussocks of love grass cleared to make small homes for these babies. The soil is hydrophobically dry and sometimes full of large river stones that are wedged out to make a deep hole for planting. Each tree is placed in its hole, filled in with dry soil, and the watering begins. Walk along to the river crossing and down the steep track, wading with buckets into the slow moving river. Scoop up buckets full of life-giving water and carry the buckets up the steep hill and along the road to the trees and fill each well with water. Over and over, each time the thirsty soil drinks deeply until several buckets later, their thirst is quenched and the trees are bedded in. Each time my heart is replenished in the soft gurgle of water over rocks, in tune with the rhythm of my own circulatory system. Every day the tiny trees speak to me, telling me what they need.

Day #2: Catastrophic fire predictions

The fires are spreading at an alarming rate, family members are in danger, the weather is expected to reach 48.7° record heat with strong north-westerly winds, the smoke grows worse. Our house may be threatened by the vast bushy ridge behind us. I need to go to the river and water the trees before it gets too hot. The wells are already dry, I repeat the ritual of walking with buckets to the river, scooping up the life-giving water and filling the wells around the tiny trees over and over until the thirsty ground is saturated. It's still hot when later that afternoon I return to find that the first two trees are burnt. The soft grey green leaves at the tips of the red box have turned a dry crackly brown and my heart bleeds as I give it another drink. The fine green leaves of the yellow box, are mottled brown all the way down and I wonder if it will survive.

Day #3 New fires start and old ones continue to spread out of control

Heart heavy and afraid, I walk out again wondering how these tiny signs of hope will be. All of the wells are bone dry. I gaze deeply into their leaves, three of them are lush rich green, so soft, so beautiful, and my heart swells with love for them. The other two show no signs of recovery but they are no worse than they were last time we communicated. They still have the dry brown patches where the leaves have died from the heat. So I water and water them all, walking up and down to the river, dipping and filling, carrying two 10-kilo buckets of water up the steep rutted track, getting to know every stone and hollow in the ground as I trudge up and down. There is something deeply satisfying about this hard labour, a relief from the endless grief, fear, anger and sorrow of what is happening to our country.

Day #4: Tender tending loved

The girls have had to leave their house in the Blue Mountains so we are going to the river together to water the trees. I need two small buckets so I empty the compost buckets from under the sink and we walk down to the river together, the same way we have often walked in happier times. They chatter all the way. At the river I show them how to dip and lift their buckets as they wade in over the river stones. Charmaine is quick and confident, Luna worries about the water weeds that slip into hers. They won't hurt the trees I tell her, but she scoops them out one by one, such is her level of care. We repeat this three times up and down, up and down, and talk about the fire fighters who died, how these trees will grow tall and strong so that we can always remember the fathers who died saving others' lives and homes. We talk about how burnt trees in the bush will one day grow, regenerate, and what regenerate means, to be born again. Young girls learning compassion, love and hope.

Day #5 Signs of new growth

I set off this morning with a sad heart, still worried about the burnt leaves and how the tiny trees do not seem to be able to recover. I come first to the last tree planted, an endangered grey box, the smallest, almost hidden in its tree guard but with a flourish of rich blue green leaves bursting out of the top. It's dry again and will need a water. The second is a taller yellow box and flings its fine leaves out of the guard, all bright green undamaged. The third is another grey box, grey green leaves splayed out and waiting for water. The fourth is the damaged yellow box, and, peering into its damaged leaves, my heart leaps with joy as I see some tiny new shoots emerging from the ends of its fine branches. Finally, I reach the first planted tree, the red box, and again I can just see the smallest signs of new shoots appearing as life and hope return.

Day #6 The rains comes

I wake to the sound of rain and walk out singing, 'just walking in the rain', singing and singing as I wade through fields of puddles lying across Emu Green. Waters roars as Lapstone Creek rushes through the drainway, bursting its banks with its wet overflow of water and debris. The trees will not need watering today, but I walk faster and faster to get to their place in the riverlands. I hurry first to the two burnt trees and whoop and yell with excitement as I notice that the tiny shoots on the yellow box have grown large and red, bursting out with new life from its tips. The red box is a more quiet revival but, sure enough, its new shoots are expanding, leaning upwards towards the sky as it grows towards its 25-metre height. The rain continues for days, sometimes a little, sometimes a torrent from a thunderstorm and each day the tiny trees produce more and more new growing shoots. All of them rejoicing in the life-giving waters that have finally come.

Others all over the firegrounds are noticing the emergence of new life, green leaves bursting from the lignotubers on the trunks of burnt trees, green grass shooting up from burnt ground, but the fires are still burning, with 30 fires still uncontained in NSW. There will be a long massive recovery effort for our bushlands and all of the living beings that depend on them, including their human communities. As Australia regenerates, there is a renewed depth of understanding of our integration within our complex local ecologies and within the global ecosystems of the planet as the smoke from our fires continues to circle the globe and plumes hover far above planet Earth.

Reference

Bodkin, F. (2017). *D'harawal: Climate and Natural Resources.* Sussex Inlet, NSW: Envirobook, Trekaway Pty Ltd.

Index

Page numbers in italics refer to figures.

Aboriginal heritage 74, 187, 188
Aboriginal people 43, 56–62, 67, 74, 78–79, 82, 121, 168–169
acacias 167, 172, 195
Afinson, John 35
Agassiz, Lake 41
AILA (Australian Institute of Landscape Architects) 70
air 13, 16, 17, 21, 41, 86, 87; and bedrocks 90, 105, 109–115; and bird song 114–115; and blue sky 90, 111–113; ever-changing qualities of 113–114; flow, and climatic warming 41; flower's fragrance 174; in-between space 113–114; poem 113; smoke 194; walking in 80, 89
allopatric speciation 5
Amazon forest 39, 176
Amazon River 176; climate entwinement 39–40; dwelling in intertwined intimacy and immensity 40; glacial source of 41; global markets, soy production and deforestation entwinement 39; poem 38; Ribereñs, entwinement of 38; socio-ecological entwinement 38–39; *see also* life and death
ancestral connections 61–63
ancestral creators 69–70, 189
animals 9, 154; becomings-animal 18–19, 20, 97, 98, 115–120, 124; and bushcare 162; habitat 8–9, 35; impact of urban development 10–11; movement of 7–8; Mulgury 68; responsibility for 68; riparian zone 8, 9
Anthropocene 40, 176; bower 11–12, 13–14; and glacial sources 41; mattering of human subject 13; mattering of life and death 12–13; mattering of time 12; walking in 14
Arden, Jacinda 179
art: and chaos 3; dragon's artworks 118; and litter 151, 155–156; and Mississippi River 34, 35; and natural selection 149–150; ochre 78–79; and songlines 65, 74
Aswan Dam 26, 27
Australian indigo 95, 160, 173
Australian Institute of Landscape Architects (AILA) 70
Austral indigo 173
Avon River 61

baiji dolphin 31, 32
Bakonjo people 42
Ballangarry, Martin 68–69
Barad, K. 107, 109, 111, 112, 142, 156
bass, and the bush 161
Bass Fishers of Sydney 23, 95, 158, 161–166, 170
bass fishing 95, 161, 166
becoming-animal 18–19, 20, 97, 98, 115–120, 124
becoming-imperceptible 18, 115–120
becoming(s): blocs of 20; blue literature review 17–20; molecular 116–117; multi-elemental becomings 124; riverlands' becomings, points of emergence 87; sacra-elemental becomings with silver and gold 110–111; songlines of 88; with the spectral 88; time of 19; writing as 20; *see also* multimodal becomings; waterways as places of becoming

becoming with water 48; choppy water memory 49; dark night 50; dawn light 49; dolls 53–54; finding home 51, 52–53; memory walks 50, 51–53; new baby 51; noon 52–53; sunny morning 51–52; water and light memories 49; water dreaming memory, stillness of 53; water memory journey 48–49; watery baby memories, closure of 54
Bedrock 89, 105; affect 90; air, and bird song 114–115; air, space in-between 113–114; as an extended reverie 106–107; becoming animal, becoming imperceptible 115–120; and birds 119–120; blue sky and air 111–113; casuarina's story 90, 121–122; colours of 90; deep time sacramental knowings 121–122; dinosaurs 119; dragon's artworks 118; eastern water dragon's dreaming 117–118; images, as play of water and light 90; multimodal becomings 90; poem 89; river text sensory ecologies 89–90; rock and stone 107–108; sacramental birds 119–120; sacramental elements 107–111, 114–115, 119–120; silver and gold 110–111; spectral 120–122; swansong 120; watery sacramental 108–110; *see also* Ganges River
beginning and ending with water: community update, Water NSW 186–191; Mississippi floods 185–186; *see also* life and death
Belloon Pass (*Belonong*) 190
biden pilosa 171; healing power of 172
Billagoola Waterhole 190
biodiversity 5, 14, 39, 42, 161, 162, 174, 182
birds 162; Bedrock 89–90, 114; Bedrock's sacramental birds 119–120; bowerbird 2–3, 84; and dinosaurs 119; emus 139; habitat 8, 35; lyre bird 147–148; sand birds 125–127; songbirds 147–148; stone stories 102
bird song 102–103, 114–115
Birrimbunnungalai Waterhole 189, 190
Birrugan, story of 69, 70
blue literature review 1; Anthropocene 11–14; chaos 3; poem 1–2; riverlands 4–11; satin bowerbird study 2–3; walking and stopping 14–17; waterways and becoming 17–20
Blue Mountains 72, 187, 197
blue sky 90, 106; and air 111–113; and Bedrocks 122
bodies: dissolving boundaries of 106–107, 108, 109; memory 49, 51, 52, 54, 67; movement and walking 15–17
bodies of water 47, 49, 80–81
bodies without organs: litter's global materialities 143–144, 153; movement of 15–17; walking 15–16
Bodkin, Francis 84, 195–196
Bogue, Ronald 150
Boorooberongal clan 57–58, 72, 74
boulders 98, 137, 140
boundaries 4; of body 106–107, 108, 109; ecotones 4–5; edge effects 5; halocline 47
bower 1–2, *2*, 84; Anthropocene 11–12, 13–14; riverlands 4, 11; of satin bowerbird 2–3; walking 14, 17; waterways as places of becoming 17
Bradley sisters 161
Bradley, William 57
Broken Bay 21, 48, 52–54, 56, 57, 59, 60, 63
Bungaree, story of 60
burial in the river's lands 177, 178, 185
Burragoorang Valley 65, 194
Burragorang Valley 61, 62, 63, 186–187, 188, 190, 192
Bushcare movement 159; bass and the bush 161; building community and tender intimacies 165–166; pace of 162–163; perpetual circle 162; planting and learning plants language 163–164; site 161, 169; weeding transformations 164
bush regeneration 23, 95, 159; riparian zone 158–159, 159–160; *see also* Bushcare movement

Callistemon salignus 173
carbon dioxide emissions 12
casuarinas 22, 59, 89, 90, 96, 133, 140; at Bedrock 89, 90, 106, 121–122; needles 96, 137, 170, 184, 192; poem 166–167; resting and nesting 167, 172; and sticks 135; story of 121–122, 137–139
Cataract River 61
catastrophic fire predictions 197
chaos of the earth 3
Children's Climate Strike 179
China: environmental protection in 32; and Nile region's hydropower projects 26
choppy water memory 49, 51, 112

Christchurch mosque shooting 179
cigarette packet, litter 142–144
Clematis aristata 173–174
climate change 12; anthropogenic 40; catastrophic 39–40; and glacial sources 31, 41
climate policy 194
Colebrook, C. 42
collection of blue 1–2, *2*, 84
colorbonds 91, 150, 152, 153
colours 96; of Bedrock 90; of drainway 100; of flowers 180, 181–182; of in-between spaces 101–102; of litter spot 92, 93; of new swim hole 94; of regeneration 96; of river's crossings 98–100; of sunlight 112, 177–178
community building 165–166
community update, Water NSW 186–187; desecrating Indigenous culture 187; 'Give a Dam Campaign' 187, 188; songline lives deep within the valley 188–191
Confidence Man, The (Melville) 34
Cordeax River 61
corridors 7, 10, 84, 160
Cox, Brian 47
creation ancestors *see* ancestral creators
Crinall, Sarah 151
Cryptozoic era 77
Cumberland Plain 77, 81, 144
Cunnark Waterhole 190
cycads 82

dams: Aswan Dam 26, 27; Gibe III dam 26; Give a Dam campaign 187, 188; Kajbar Dam 26; Karuma Dam 26; Nile River 26, 27; Three Gorges Dam 30; Warragamba Dam 43, 61, 186, 187, 188, 192
Darug Country 135; Bedrocks 121; contemporary songlines 70; culture 43, 74; heritage 74; people of 21, 65, 80, 135; songlines 43, 65, 72, 74, 79; Yellomundee Regional Park 80
Davey, Wee, story of 61–62, *63*, 192
death *see* life and death; spot of death
deep listening, stopping to 16, 17, 156
Deerubin 56
deforestation 39–40
Degeorges, P. 176, 178, 182
Deleuze, G. 17, 40–41, 116–117, 118, 192
destruction: and regeneration 170–171; and Yangtze River 30–32
Dharawal people, string making 138–139, 140
Dharawal season of *Ngoonungi, Murrai'yunggory* 172
dinosaurs, avian and non-avian 119
Diprotodon 82
Djuli of the Mudong cycle 195
dragons 118–119; artworks of 118; eastern water dragon's dreaming 117–118; water dragons 93, 115
drainway spot 99–100; colours of 100; and flowers 178–179; multimodal becomings of 100; song of 98–99
drawing 3, 58, 134–135
dream analysis 88–89, 101
dreaming 78, 117–118
dreaming memory, water 53
dreams and possibilities, river's litter 145–146
dreamtime stories 68, 78, 118, 140, 187, 189
drought 42, 47, 93, 160, 186
ducks, at river's crossings 131

eastern water dragon's dreaming 117–118
ecotones 4–5, 11; and edge effects 5; riparian zones as 6; and speciation 5–6
edges, definition of 5
eel story, river's crossing 136–137
Egypt: Nile River 26; water myth 27
Emu Green riverlands 4, 20, 43, 46. 86, 99–100, 139–140, 142–155
environmental volunteering programmes 159, 161, 165
ethical walking 15, 17
eucalypts 79–80, 96, 171, 195–196
Eucalyptus melliodora 196, 197, 198
Eucalyptus microcarpa 196, 198
Eucalyptus polyanthemos 196, 197, 198
Eucalyptus tereticornis 160
European Union: Water Framework Directive (WFD) 36
ever-present river 83–84
extinction 12–13, 14, 23, 31, 40, 42, 82, 119, 176

fire 194; catastrophic fire predictions 197; environment prone to 160; and regeneration 197; regimes 79–80; wildfires 195–196
fish 20, 31 37, 43, 59; *see also* fishing
fishing 38, 161; Bass Fishers of Sydney 23, 95, 158, 161–166, 170; bass fishing 95, 161, 166; regeneration spot 158
fishing spot 91, 147

floods/flooding 8, 97, 103, 146, 154, 156, 192; impact of 7; Mississippi River 185–186; and mud 128, 130, 140; Nile River 27; Rhine River 42; and vegetation 6; Warragamba River 186–191; Yangtze River 31
flowers: colours of 180, 181–182; flower-laying ceremony 180, 181–182
flowers, and bush care 172; green wattle 172–173; *Indigofera* 173; white bottlebrush 173; white cedar 174; white clematis 173–174
flowers of eternal life: flower-laying ceremony 180, 181–182; flowers for massacre's trauma 179–180; harvesting flowers for ceremony 180–181; poem 178–179
fossils 12, 13, 119
Foucault's fossils 13
fresh water 6, 42, 47–48; and salt waters, meeting of 21, 46, 47–48, 54, 56, 58–60, 63
friction 15, 17, 98, 185

gal-gal-la (smallpox) 58–59
Ganges River 105; bacteriophage presence in 29; basin 28–29; Gangajal 29; glacial source of 41; materiality of 28–29; poem 28; pollution of 29; sacramental materialities 29–30; sacredness 28; sacredness, materiality of 29; *see also* Bedrock
Gareem 190
Gaumukh 41
Genyornis 82
'Ghost Trails' (sculpture) 34
Gibe III dam 26
Give a Dam campaign 187, 188
glacial sources 40–41, 122; and Anthropocene 41; and climate change 41; of Ganges River 41; of Mississippi River 41; of Nile River 42; of Rhine River 42; of Yangtze River 41–42
glaciers, decreasing of 41–42
global change, effects of 5–6
gold, Bedrock's 110–111
golden cloak (symbolising trauma) 182; binding of 182–185; burial hole for 184–185
Gomuck 122
Gomukh 29
Gondwanaland 43, 79, 160, 194
Goorit Waterhole 190
grass 17, 20, 168–169
grasslands 101–102, 132, 139
Great Dying 12
Great Ice Age 41
green wattle 102, 172–173
Grosz, Elizabeth 3, 115, 149–150
Guattari, F. 17, 116–117, 118
Gudalong Marool 195, 196
Gumbaynggirr Country: contemporary songlines in 68–69; songlines as walking trails of ancestral creators 69–70
Gundungurra Aboriginal Heritage Association 189, 190
Gundungurra Country 188, 190; Dreamtime stories 187; songline 187, 189; water's beginning 61
Gunggalook Waterhole 190
Gurangatch: resting/hiding place of 190

halocline 47
Hawkesbury Nepean River 43, 46, 48, 62, 72; flood risk 186, 187; songlines 65, 80, 84; watery ways 86
Hawkesbury River 43, 72, 80, waters of 56–57; Sandstone map 77
Henderson Island 22, 35, 141, 145, 146, 154, 155, 156
hoons, stories of 170
Huckleberry Finn (Twain) 34
human activity, and changes in Earth system 12
human enmeshments 10–11
human subject 13, 40–41
hunter-gatherers 80
Hunter, J. 57
hydrothermal vent 47–48

images: of Bedrock 90, 110; of dragon 118, 119; dream analysis 88–89, 101; in-between space 101, 103; of litter spots 92, 143; of regeneration 96; of river's crossing 98; songlines 76–80, 88; of stone 102; tattoo images 134
in-between spaces 88, 101–103; air 113–114; bird song 102–103; colours of 101–102; grasslands 101–102; river's weaving 103; stone stories 102; water's songs 103
Indigenous culture, desecration of 187
indigoferas 167, 173
indiscernibility zones 18, 20
inhuman 15, 22, 86, 141, 142
insects 84, 160, 162, 166; and regeneration 173, 174; vitality of 9

intra-action 15, 17, 98, 142
invertebrates 9–10

Jacobson, Susan 151
Jones, L. 112–113
Joolundoo Waterhole 189, 190

Kajbar Dam 26
kangaroo grass project 168–169
Karskens, G. 57–59, 190
Karuma Dam 26
Kenya: Nile River 26
knowing and learning 150
Kumar, D. 28

language: of plants 163–164, 168; of river 146; of stones 102, 107; of waters 146, 189
Lapstone monocline 72, 77, 80, 81, 194
Laurasia 79
life: flowers of eternal life 179–182; memory traces of everyday life 52, 53; and rivers 25–26, 28, 30, 33, 35, 38; water's beginnings of 47–48; *see also* regeneration
life and death 101; mattering of 12–13; song of 100–101; in the tree world 96; on water 58–59; on the waters 58–59; *see also* Amazon River; spot of death
light: blue sky 112–113; dark night 50; dawn light 49, 54–55; sunlight 49, 50, 90, 111–112, 177–178, 184; walking in 80–83; and water 49, *50*, 90
litter 141, 162; affect of 92; and art 151, 155–156; artful proliferation of 155–156; colours of 92, 93; communications of 142–143; continuing global connections 145; dreams and possibilities 145–146; global materialities of 143–144, 153; Hendersen Island 145; installation creation 151–153, 155–156; knowing and learning 150–151; language of 147; movement of trash 154–155; moving with the flow of 151–154; multimodal becomings of 92; orange chairs' story 147–148; poem 141–142, 147, 148, 149, 150–151, 154, 155; pristine beach and stones 142; song of 91; struggle with 149–150; texts 91–92; water flows and force expansion 148–149; wild thinking with water 145–146; *see also* Mississippi River
living matter 13, 185
lizard, and becoming animal 116–117
local places 14, 91, 158
Lock, Maria 70, 71, *71*, 72, 73–74
Lorelei, legend of 37
loss, dialogue with 14, 17
Luigi di Savoia, the Duke of Abruzzi 42
Lu, X.X. 30, 31
lyre bird 147–148

macro-invertebrates 9
macrozamia 80, 82
making community: destruction and regeneration 170–171; healing power of biden pilosa 172; planting and seeding 167–168; poem 166–167; prickly intimacies 171; resting, nesting, and casuarina tree 167; tender intimacies 170; watering 169–170; weed's regenerative powers 171
Marshall, Chrissiejoy (Immiboagurra-milbun) 67–68
marsupials 82, 131–133, 135
massacre's trauma, flowers for 179–180
mass extinctions 12–13
material flows 7
Mato Grosso 39
mattering: of human subject 13; of life and death 12–13; of time 12
megafauna 82
Melia azedarach 174
Melville, Herman 34
memories: body memory 49, 51, 52, 54, 67; choppy water memory 49; memory traces of everyday life 52, 53; memory walks 50, 51–53; of molecule 20; of a molecule 20; water and light memories 49; water dreaming memory, stillness of 53; water memory journey 48–49
meristematic tissue, regeneration 160
Merowe reservoir 26
microbiota 9
Miller, Angela 34
Mississippi River: art creation 34; blues music 33–34; creating with 35; floods 185–186; glacial source of 41; and music creation 33–34; poem 33; stories creation 34–35; *see also* litter
Modes, Wes 34
molecule, memories of 20
moment, the 20, 87
more-than-human 4, 11, 14, 15, 17, 48
Morle Boc 190
movement: and friction 15, 185; of litter 143, 144, 145, 154, 155, 156; new

swim hole 92, 93; of plants and animals 7–8, 174; and rest 116, 117; of water's endless flow 32, 49, 81; *see also* walking
moving in time and space 17
mud girls 128–131
Mudong cycle 195
Mulgury (totem) 67–68
multi-elemental becomings 124
multimodal becomings: of bedrocks 90; of drainway 100; of litter spot 92; of new swim hole 94; of regeneration 96; of river's crossings 98
multimodal images 88
multispecies 9, 124
Murraural Waterhole 190

Nannup, Noel 121–122
Narran lake 67
Nepean River 4, 61, 84; upper canal system 61; waters 57–58; *see also* Hawkesbury Nepean River
Nepean Trail 76
nesting 8, 9, 167, 188, 196
new growth 39, 95, 160, 198
new swim hole: colours of 94; multimodal becomings of 94; song of 92–93
New Zealand: Christchurch mosque shootings 179–180
Ngoonungi, Murrai'yunggory 172, 174
ngurra 79–80, 83, 91, 134
Nile River: dams 26, 27; exploration of the lost 27; fluvial myths of 27; glacial source of 42; hydropolitics 26; poem 25–26; touching and being touched by 27; water conflicts 26; *see also* songlines
Noongahburrah people 68
NSW Department of Planning, Industry and Environment 189

ochre: as art material 78–79; creation story of 78
Open Rivers (Nunnally) 34
Oppermann, S. 176, 178, 182
orange chairs, story of 103, 147–148

Palaeozoic era 12
Pangea 79
parapatric speciation 5
Phanerozoic era 77
Phillip, Arthur 56, 57, 58, 72
photologs 88
Pintubi Luritja women 66–67, 79
Pintupi Luritja women *66*

Piotrowski, M. 114
places, and histories 69
planting 163–164, 167–168, 196–197
plants, movement of 7–8
plastic rubbish 145, 154, 155; *see also* litter
plutonium fallout 12
poem: Bedrocks 107, 108–109, 110, 114, 115, 119, 120–121; bird song 102; blue sky 111; bodies of water 80–81; casuarina tree 137–138, 166–167; drainway 178–179; grasslands 101; island home 131–132; macrozamia 82; mud girls 128, 129; river is always there 65; riverlands 18–19, 19–20; rivers and life (Amazon) 38; rivers and life (Ganges) 28; rivers and life (Mississippi) 33; rivers and life (Nile) 25–26; rivers and life (Rhine) 35; rivers and life (Yangtze) 30; river's crossings 124–125; river's litter 141–142, 147, 148, 149, 150–151, 154, 155; sand birds 125–127; spot of death 183, 184; stone 102, 108; swim hole 46; waiting 133–134; water's songs 103; writing on water 16–17
poepatetics 16–17
pollination 9, 96, 172, 173, 174
pollution 10, 29, 36, 40, 144, 153; *see also* litter
posthumanism 13, 14, 43, 185
prickles 171
pristine beach and stones 142
pure event, time of 19–20

rains 146, 198–199
regeneration 163, 199; colours of 96; flowers and bush care 172–174; making community 166–172; multimodal becomings of 96; New Zealand's one billion trees plantation goal 158; poem 160; site 160–161; song of 94–95; trees' power of 194; waterways transformation 174; *see also* Bushcare movement; bush regeneration; life and death; Rhine River; *see also*
replantation 95, 163, 164
reproduction 9
reptiles 8, 93–94, 116–117, 118
resting 57–58, 167, 190
restoration of natural area 162–163
retention process 163
Rhine River 159; basin 36; glacial source of 42; mythical transformations

(Lorelei) 37; poem 35; policy processes transformation 36; pristine to industrial transformation 36; regeneration 37; uncommon transformation 36; Water Spirit temptress to Forsaken Bride transformation 37; *see also* regeneration
Ribereños 38
riparian zones 11, 31; bushcare regeneration 158, 159–160; as ecotone 6; habitat 8–9, 10; human enmeshments 10–11; and material flows 7; as most productive eco-systems on Earth 8; and movement of plants and animals 7–8; multispecies and interdependence 9–10; transformation of 174
riverlands: bower 4, 11; ecotones 4–5; edge effects 5; habitat 8–9; human enmeshments 10–11; material flows 7; movement of plants and animals 7–8; multispecies and interdependence 9–10; riparian zones 8–11; riparian zones as ecotone 6; transcorporeal bodies 11
riverlands' becomings, points of emergence 87
riverlands' watery ways 86; bedrock 89–90; drainway spot 98–100; litter spot 91–92; moments of encounter 87; new swim hole 92–94; regeneration spot 94–96; river's crossing spot 96–98; space of the in-between 101–103; spot of death 100–101; walking the book's songlines 87–89; *see also* Hawkesbury Nepean River
river red gum 77, 79, 83
river's book, beginning and end of 101, 177
river's crossings 124–125; casuarina's story 137–139; colours of 98–100; Dadda rocks, Mama rocks and bubba kangaroos 131–132; ducks 131; eel story 136–137; mud girls 128–131; multimodal becomings of 98; the other side 139–140; poem 124–125, 125–127; river camp tattoo parlour and teddies 133–135; river stones tatts 134–135; rock jagwars 137; sand birds 125–127; seal island 136–137; song of 96–97; spot of 97; sticks 135; swamp wallaby 132–133; turtle's story 127–128; wombat story 135; *see also* Yangtze River
river's litter *see* litter
river's sacramental 105–106, 107
river stones: and flowers 178; tatts 134–135
river's weaving 103
rock and stone: of Bedrock 107–108; Dadda rocks, Mama rocks and bubba kangaroos 131–132; river stones tatts 134–135; rock jagwars 137; sandstone 76–80; stone stories 102
rubbish *see* litter
Rwenzoris Mountains 42

sacra-elemental becomings with silver and gold 110–111
sacramentals 105–106, 107
salt water 47–48, 53, 79; and fresh waters, meeting of 21, 46, 47–48, 54, 56, 58–60, 63
sand birds 125–127
Sandoz accident (1986) 36
sandstone 77; creation story of 77; walking in 76–80
satin bowerbird 2–3
Schama, Simon 27
Schneiderman, J.S. 186
Scotland Island 46, 48, 54–55, 56, 60, 62
seal island 98, 135, 136–137
seasons 113–114; change of 113–114; *Ngoonungi*, *Murrai'yunggory* 172, 174; of regeneration 94–95, 196
seasons of planting and regeneration of hope: catastrophic fire predictions 197; New fires start and old ones continue to spread out of control 197; planting day 196–197; rains 198–199; signs of new growth 198; Tender tending loved 197–198
'Secret History' project (Modes) 34
sediments/sedimentary rocks 77–78
seeding 168–169, 172
self 17–18, 110, 116
shells 59
Shiva, Vandana 26, 27
silver, Bedrock's 110
singing painting river 80
sitting 16, 69
skin name 66–67
sky 132, 178; *see also* blue sky
smallpox epidemic 58
smoke 194
smoking 143–144
snapper and bream 59
soil 6, 7, 8, 159, 168, 197
songbirds 147–148
songlines 65; of becoming 88; contemporary, in Darug Country 70;

contemporary, in Gumbaynggirr country 68–69; *Mulgury* as totemic relation 67–68; outset 66–67; singing and dancing in ceremony 79; *Song Sings On* 75–76; swans as Mulgury 68; and totemic connections 67; walking the singing painting river 76–83; as walking trails of ancestral creators 69–70; *Wirrawee* 73–74; of Yarramundi 72–73; *see also* Nile River
Song Sings On 75–76, 98–99
sound of river 129
Southern Cross 70
space *see* in-between spaces)
spatio-temporal assemblages of events 19
speciation 5–6
spectral, becoming with 88
Spence, L. 37
spot methodology 86, 87; *see also* Bedrock; drainway spot; litter; new swim hole; regeneration; river's crossings; spot of death
spot of death: beginning and end of the river's book 177; burial in the river's lands 177; and colours 177–178; poem 176–177, 183, 184; small girls, river stones and flowers 178; song of 100–101; *see also* life and death
sticks, river's crossing 135
stone *see* rock and stone
stopping 16
stream water 6
sunlight 49, 50, 90, 111–112, 177–178, 184
swamp wallaby 132–133
swans, as totem 68
Sydney 60–61; ancestral connections circle, closure of 61–63; Bass Fishers 23, 95, 158, 161–166, 170; water shortages, Gundungurra, and Warragamba Dam 61
sympatric speciation 5

Tank Stream 60
tattoo and tattoo parlour 134–135
tender intimacies 165–166, 170
Terewah Country 67–68
terrestrial animals 9
territory, of sand birds 126–127; *see also* boundaries
tetom 67–68
Thomson, Andrew 59, 60
thought experiments 156
Three Gorges Dam 30, 31, 32; *see also* Yangtze River
Tian-e-Zhou Oxbow Reserve 32
time: matter of 12; of pure event or of becoming 19–20
time and space, walking as embodiment of 15
Tobin, Jacinta and Leanne 70, 74, 99; and Darug heritage 74; Leanne's paintings 71, 79; song, on Yarramundi 72; *Song Sings On* 75–76; *Wirrawee* 74
Tongtian River 41
touch/touching 27, 67, 108
transcorporeal bodies 11, 13
trees plantation 158, 162
trees, watering of 169–170
turtle's story, river's crossings 127–128
Twain, Mark 34

Uganda: Nile River 26

Varanasi 29
videos 76, 77, 80, 86, 88–89, 105, 169

walking 50–53; in air, light, water and strange red fruit 80–83; in Anthropocene 14; bower 14, 17; dialogue with loss 14; as embodiment of time and space 15; ethical walking 15, 17; as methodology 15; as moving body-without-organs 15–16; with multiple others 15; poepatetics 16–17; in sandstone and ochre 76–80; and stopping for deep listening 16–17; trails 69–70; traversing terrains of world and mind 16
walking the book's songlines 87–89; becoming with the spectral 88; dream analysis 88–89; multimodal images 88; poem 87–88; songlines of becoming 88
wallaby 132–133
Waratah Rock 190
Warragamba Dam 43, 61, 186, 187, 188, 192
Warrumba 190
waste 12; *see also* litter
water dragon: artworks 118; becoming imperceptible with 118; dreaming 117–118
watering trees 169–170
waters 156, 194; alchemy, and creation 81; becoming with 48–54; beginning and ending with 185–191; beginnings of life on Earth 47–48; bodies of 47, 49, 80–81; choppy water memory 49; conflict, Nile River 26; dreaming

memory 53; fresh and salt waters, meeting of 21, 46, 47–48, 54, 56, 58–60, 63; freshwater 6, 42, 47–48; life and death on 58–59; and light memories 49, *50*, 90; memory journey 48–49, 54; memory traces 57; and mud 129; sacramental 108–110; saltwater 47–48, 53, 79; songs of 103; and Sydney's growth story 60–63; wild thinking with 145–146; world-becoming with 47–48
waters, of first contact 56; Hawkesbury River waters 56–57; land plundering 59; naming and claiming of land 59–60; Nepean waters and Boorooberongal clan 57–58
waters, of migration: material connections 54–55; Scotland Island, dawn of new day 54–55
Water Wars (Shiva) 26
waterways as places of becoming 6; becoming-animal becoming imperceptible 18–19; blocs of becoming 20; bower 17; encounters of becoming 17–18; memories of molecule 20; spatio-temporal assemblages of events as encounters 19; time of pure event or of becoming 19–20; writing as becoming 20
wattles 84, 94, 102, 167, 172–173, 195
weeding transformations 164–165
weeds 162, 171
white bottlebrush 173
white cedar 174
white clematis 173–174
wildfires 195, 195–196
Williams, Bernard 34
willow bottlebrush 173
Wirrawee, walking with 73–74
wombat story 135
Wonggaree (Slippery Rock) 189
Woonggaree Waterhole 190
world-becoming with water 47–48
World Wildlife Fund (WWW) 32
writing, as becoming 20

Yangtze River 30, 31; *baiji* dolphin 32; *baiji* dolphin extinction 31; dams 30; declination of finless porpoises 31; destructions 30–31, 32; finless porpoise 32; glacial source of 41–42; life force of 32; poem 30; Princess Baiji, story of 31; salvations 30, 31, 32; *see also* river's crossings
Yarramundi 72–73; Leanne and Jacinta's song on 72–73; songline 80
Yellomundee 80
Yellomundee Regional Park 76, 80

Zalasiewicz, J. 40

For Product Safety Concerns and Information please contact our EU representative GPSR@taylorandfrancis.com
Taylor & Francis Verlag GmbH, Kaufingerstraße 24, 80331 München, Germany

www.ingramcontent.com/pod-product-compliance
Ingram Content Group UK Ltd.
Pitfield, Milton Keynes, MK11 3LW, UK
UKHW011008180926
475701UK00004B/634

9780367497835